ABOUT ISLAND PRESS

Island Press is the only nonprofit organization in the United States whose principal purpose is the publication of books on environmental issues and natural resource management. We provide solutions-oriented information to professionals, public officials, business and community leaders, and concerned citizens who are shaping responses to environmental problems.

Since 1984, Island Press has been the leading provider of timely and practical books that take a multidisciplinary approach to critical environmental concerns. Our growing list of titles reflects our commitment to bringing the best of an expanding body of literature to the environmental community throughout North America and the world.

Support for Island Press is provided by the Agua Fund, The Geraldine R. Dodge Foundation, Doris Duke Charitable Foundation, The Ford Foundation, The William and Flora Hewlett Foundation, The Joyce Foundation, Kendeda Sustainability Fund of the Tides Foundation, The Forrest & Frances Lattner Foundation, The Henry Luce Foundation, The John D. and Catherine T. MacArthur Foundation, The Marisla Foundation, The Andrew W. Mellon Foundation, Gordon and Betty Moore Foundation, The Curtis and Edith Munson Foundation, Oak Foundation, The Overbrook Foundation, The David and Lucile Packard Foundation, Wallace Global Fund, The Winslow Foundation, and other generous donors.

The opinions expressed in this book are those of the author(s) and do not necessarily reflect the views of these foundations.

The Conservation Professional's Guide to Working with People

The Conservation Professional's Guide to Working with People

Scott A. Bonar

Foreword by Duane L. Shroufe

ISLANDPRESS Washington • Covelo • London

Library of Congress Cataloging-in-Publication Data

Bonar, Scott A.
The conservation professional's guide to working with people / Scott A. Bonar; foreword by Duane Shroufe.
    p. cm.
Includes bibliographical references.
ISBN-13: 978-1-59726-147-0 (cloth : alk. paper)
ISBN-10: 1-59726-147-5 (cloth : alk. paper)
ISBN-13: 978-1-59726-148-7 (pbk. : alk. paper)
ISBN-10: 1-59726-148-3 (pbk. : alk. paper)
  1. Communication in conservation of natural resources. 2. Conservation of natural resources—Public relations. 3.Interpersonal communication. 4. Conservation of natural resources—United States. 5. Natural resources—Management. 6. Natural resources—United States—Management. I. Title.
  S944.53.C65B66 2007
  333.7201'4--dc22
                                        2006037237

Printed on recycled, acid-free paper ♻

Manufactured in the United States of America

10 9 8 7 6 5 4 3 2 1

For John and Dorothy, who taught me best about the importance of a healthy environment and good people skills.

For Sophia and Sonja, my two best reasons for helping to protect the environment.

CONTENTS

FOREWORD

Fortunately, the natural resources profession continues to evolve, albeit slowly. Just as organisms in natural systems must evolve to survive in changing environments, so too must we as professionals responsible for public trust resources evolve to address new challenges and greater expectations. In 1930, Aldo Leopold challenged the American public to recognize that law enforcement was an insufficient tool in conserving wildlife resources, bringing attention to the need for adding research and management programs to our repertoire.

Today, biological science and law enforcement, by themselves, are insufficient for managing natural resources. To effectively manage natural resources now, professionals need to develop skills in dealing with people. My profession has been perhaps too slow in understanding that wildlife management is as much a social discipline as it is a scientific one. But, this realization is finally becoming nearly universal among leaders in the profession. We manage wildlife resources for the benefit of the people who have entrusted their care to us, the stewards of a publicly owned resource. After all, elk and eagles don't set natural resources policy, people do.

Criticism from both employees and constituents about ineffective people skills abound in natural resources agencies. Conservation professionals need to communicate more effectively. Further, we need to better understand the importance of building relationships among ourselves and with all constituents. The days have long passed where we can hide behind the attitude that "we are the professionals, trust us, don't question us." The public expects and deserves accountability from our profession. We must embrace accountability, not be offended by it. Our profession must strive to satisfy customers. We are not, in any way, exempt from that motive, common in private industry. We need to continuously practice customer service as part of the fabric of who we are. Moreover, we must stretch beyond customer satisfaction and build customer relationships. We must build equity in those relationships through trust, transparency, responsiveness, and quality in our products and services. Equity in public relationships pays dividends in agency support, both socially and financially. It creates public advocacy for our agencies and programs, and allows for forgiveness when we err, which we have and will again. People empower us to do our jobs; yet they are also capable of disempowering us.

The importance of social skill in the natural resources profession is constantly growing. Public expectation and scrutiny of its servants is ever increasing and its zenith is not in sight. Further, the urbanization of America has resulted in a growing percentage of the public that lacks a connection to the land and an understanding of natural systems. Competition for natural resources and a polarization of attitudes regarding their sustainable use and welfare compound our challenge. Educating a sometimes-misinformed public and building consensus among deeply invested stakeholders of conflicting interests requires skills beyond those needed twenty years ago. Now, skills in communication, education, negotiation, and conflict resolution are as important as those in law enforcement or fisheries and wildlife science.

Generally speaking, academia has vastly improved the preparation of newly trained conservation professionals in understanding the importance of being socially effective. Yet, there remains room for improvement. Scott Bonar has made a significant contribution in filling this need with *The Conservation Professional's Guide to Working with People*. This text should be read and practiced by new and tenured professionals alike. Scott is uniquely qualified in offering guidance on this topic. He has devoted many years to ensuring his students enter our profession prepared for the biological and socio-political challenges they will face during their careers. Scott has successfully demonstrated his ability, not just in the classroom, but also in dealing effectively with the public on contentious issues. Scott has maintained a close relationship with natural resources agencies to ensure that his students are appropriately prepared to be successful in the rapidly changing profession they have chosen. For that, leaders in my profession are grateful, and indebted to Scott.

Duane L. Shroufe, Director
Arizona Game and Fish Department

PREFACE

I graduated with a Ph.D. with honors in Fisheries Science, started work at a fish and wildlife agency, and found I was unprepared. I learned quickly that successful natural resources management is much more than good science. It requires working with angry landowners, meeting deadlines, dealing with stress, supervising staff, and cooperating with politicians. However, my college background in natural resources focused on science, not the "people" skills needed to be most effective. Social skills are critical for conservation professionals, but with the exception of negotiation, few of these skills have been discussed in the context of the environmental profession. To improve my own people skills, I spent years studying effective conservation professionals and watched them closely to see what they did differently. I also studied government figures from history, including generals, politicians, administrators, diplomats, and managers, to see what makes an effective government worker. Finally, I read numerous books and articles on psychology, communication, organizational skills, sales, customer service, and stress management. By using this information, I was able to become more effective, decrease stress, and develop friendships with a wide variety of people. I have worked over twenty years in state and federal government, academia, and private industry. This book, a distillation of what I learned, is intended to help natural resources professionals work effectively with people—a critical skill for successful resource conservation.

The Conservation Professional's Guide to Working with People is a practical how-to guide of people skills for natural resources professionals. The book is designed to be both easy and interesting to read. I discuss how to increase your effectiveness using social psychology, negotiation, influence, conflict resolution, and verbal judo, managing personnel, time management, and funding techniques. Application of these skills is illustrated with examples from history, current events, and the natural resources profession to hopefully educate and entertain you.

This book should be on the shelf of environmental professionals who want to improve their "people" skills. Those who are already good at working with others will learn new tips. Those who are petrified of conducting public meetings, requesting funding, or working with constituents will find easy, common-sense advice about how to begin.

No book, especially one on people skills, is written without considerable help from others. I really appreciate the help of many talented supervisors, teachers, coworkers, students, and employees that helped me learn and apply people skills. I especially thank Lee Blankenship, Bruce Bolding, Craig Burley, Bruce Crawford, Penny Cusick, Marc Divens, Doug Fletcher, Ross Fuller, Robert Gibbons, Molly Hallock, Terry Jackson, Rich Lincoln, William Meyer, Paul Mongillo, Steve Schroeder, Jim Scott, Scott Smith, and William Zook from the Washington Department of Fish and Wildlife; Kathy Hamel from the Washington Department of Ecology; Jim Fleming, M. Lynn Haines, Bernard Shanks, and Ken Williams from the USGS Cooperative Units Research Program Headquarters, Rory Aikens, Rob Betasso, Scott Bryan, Jim deVos, Don Mitchell, Larry Riley, Scott Rogers, Duane Shroufe, Eric Swanson, Bruce Taubert, David Ward, Dave Weedman, and Kirk Young from the Arizona Game and Fish Department; David Beauchamp, Jonathan Frodge, Christian Grue, Gilbert B. Pauley, T. Brock Stables, Gary L. Thomas, and David Weigand from the Washington Cooperative Fish and Wildlife Research Unit; Tom Archdeacon, Cori Carveth, Courtney Conway, Melanie Culver, Alexander Didenko, Jon Flinders, Jason Kline, Yuliya Kuzmenko, Alison Iles, Anne Kretschmann, Laura Leslie, Charles Schade, Andrew Shultz, Erica Sontz, Timofy Specivy, Sean Tackley, Cora Varas, Cristina Velez, Ann Widmer, and Carol Yde from the Arizona Cooperative Fish and Wildlife Unit; Charles Ault, Paul Barrett, Sherry Barrett, Mark Brouder, Stewart Jacks, and Pam Sponholtz from the U.S. Fish and Wildlife Service; Jeff Simms from the U.S. Bureau of Land Management; Paul Krausman, William Mannan, William Matter, Patrick Reid, and William Shaw from the University of Arizona; David Willis from South Dakota State University; Wayne Hubert from the Wyoming Cooperative Fish and Wildlife Research Unit; and William Davies from Auburn University; and Peter Cinquemani from the Federal Mediation and Conciliation Service.

To write this manuscript, I benefited from the work of many psychologists, organizational specialists, and negotiators, especially David Burns,

Robert Cialdini, Roger Fisher, Lee Iacocca, Alan Lakein, Michael LeBoeuf, Paul Levy, Bruce Patton, Lawrence Susskind, Jennifer Thomas-Larmer, George Thompson, William Ury, and Stephanie Winston. Their practical methods have made interactions easier among millions of people. Scientists such as Paul R. Ehrlich and Edward O. Wilson have effectively communicated the need for us to conserve our natural resources, and their writings have given us ideas of how to live on this planet with the least impact. In this text I have tried to share their stories and techniques, as well as those from dozens of additional authors. The credit for these methods is theirs, and any mistakes in interpretation are my own.

Many people are worthy of special attention because of their hard work providing suggestions, editing, and improving this text. These include my parents, Dorothy and John Bonar, who, besides being former educators and good editors, passed on valuable people skills to me that I use every day. Ann Flaata, (Independence High School) English department chair (Glendale, Arizona), edited this document with considerable skill. David Walker, a good friend, business owner, and outstanding editor, reviewed this work so it would be applicable to those in both the private and public sector. William Shaw and Paul Krausman, world-renowned wildlife researchers from the University of Arizona, used their significant talents to help ensure the book was relevant, useful, and accurate. Alison Iles, from the University of Arizona, is especially talented in dealing with students and staff. I appreciated all her outstanding comments. Bernard Shanks, who has worked for many years in state and federal government as well as in academia, gave me valuable advice about public service and offered excellent editorial comments. Phil Pister, a retired biologist with the California Department of Fish and Game, is respected internationally for his work in fish and wildlife conservation and bioethics, and was instrumental in providing information about the Devil's Hole case, and other great suggestions for the book.

I also would like to thank my editors at Island Press, Barbara Dean and Barbara Youngblood. They provided exceptional guidance for this work under tight deadlines. I thank other staff at Island Press as well, including Todd Baldwin, Emily Davis, Jessica Heise, Erin Johnson, John Cangany, Alexander Schoenfeld, and Brian Weese. Their friendly, knowledgeable attitude and expertise helped move this project along quickly.

Finally I would like to thank my daughters, Sophia and Sonja Bonar, for their good humor, suggestions, and patience while Dad was struggling over each draft.

CHAPTER ONE

A Personal Story

Before we arrived, I knew there would probably be trouble. I was a biologist for the state of Washington and we were going to sample fish in rural Lake Alton[1] in far northwestern Washington. The far western and northern parts of Washington consist of the Olympic, Key, and Kitsap peninsulas. Washington's Olympic Peninsula was a land of deep cedar and hemlock forests, cold mist and rain, crashing surf, and ice-capped summits—a spectacularly beautiful place. The peninsula was one of the last places explored in the continental United States by Europeans. Almost nothing was known about its interior until reconnaissance expeditions led by Lt. Joseph P. O'Neil in the late 1880s hacked and pushed through the soaking, thick vegetation and across steep icy crests, blazing the first trails across this unknown land.[2] The Kitsap and Key peninsulas contained bedroom communities for Seattle on their far eastern sides, and supported two major navy bases; however, their interiors held thickets of second-growth Douglas fir, alder, blackberry, salal, and gravel roads on which you could twist and turn for hours before finding your way out.

Northwestern Washington was never densely populated, and many of those who did live there harbored a rich animosity for government officials, especially conservation professionals. Post–Vietnam era newsletters and magazine articles spoke of "tripwire veterans" dealing with post-traumatic stress syndrome who roamed the deep woods of the area, living off the land and shunning public contact.[3] The town of Aberdeen, on the southern end of the Olympic Peninsula, had the reputation of being the wildest town west of the Mississippi because of excessive gambling, violence, drug use, and prostitution. It was declared off-limits to military personnel as late as the 1980s.[4]

Rapid harvesting of timber and destruction of habitat for the northern spotted owl led the government to restrict the amount of logging that was conducted, which increased animosity even more. Residents of Forks, Washington, a small town in the heart of the Olympic Peninsula, painted all of their fire hydrants to look like loggers, and boasted a holiday called "James Watt Appreciation Day,"[5] named after Ronald Reagan's controversial secretary of the interior who was the bane of many environmental groups. At that time, towns throughout the area were known to be unfriendly, even dangerous, to those wearing the uniform of a state or federal conservation agency. Two of my friends, fisheries biologists for the state fish and wildlife department, came under rifle fire from a disgruntled citizen at a rural lake when they were conducting an electrofishing survey in the area. My racquetball partner, also an agency biologist, was beaten up by irate commercial fishermen.

The study of interactions between salmon and introduced fish, such as largemouth bass, was ranked as the highest priority for our research team by fellow state fisheries biologists, and Lake Alton was the perfect site for studying these interactions. Lake Alton was about forty acres and was surrounded by cattail marsh and a few ranch-style and two-story wooden houses. The lake had a large population of coho salmon migrating through it, and it also contained a healthy population of introduced fishes. Several miles downstream, Washington Department of Fish and Wildlife biologists had monitored the run of juvenile salmon leaving the watershed for the past twenty years, using a trap located on a small, sunny tidal flat next to Puget Sound.

We were legally entitled to sample the fishes of this lake, and had called landowners who had given us permission to launch our boats from a small common area on the lake. The lake was different than most in that there were two, not one, homeowners associations. While one homeowners association was very cooperative, the other refused all of our efforts to contact them in order to explain the purpose of our project. When I phoned and asked if I could send them some information about our project, a cold voice on the other end of the line said I could send it to them in care of "Fort Alton." They told us, in no uncertain terms, that they did not want us to do a study on "their" lake. Having little luck interacting with this group, we decided to launch our boat from the side of the lake owned by the friendly homeowners association.

On a cold, damp April evening we drove to the lake to launch our electrofishing boat from the common area and sample the fish populations. I led the crew, which consisted of two other biologists and a technician. Large Douglas fir and western hemlock trees lined the twisting two-lane road, small puddles soaked the black pavement, and little clouds of white mist marched across the darkening adjacent hills. The wipers clacked from side to side, and we had the air conditioner turned on to high heat to suck the excess moisture from the windshield so we could see out.

As we neared the lake, a dirty, white, late-model pickup truck appeared in my sideview mirror. It followed closely and would not pass. I felt my stomach turn uneasily as it became apparent that the truck was not trying to get somewhere, but was slowly following us. As we neared the launch, we pulled our trucks to the side of the road and stopped to ready the boat for the lake. Then the white truck gunned around the front of our truck and pulled in at an angle, blocking our way. It screeched to a stop, and a stocky man, dressed in a white, short-sleeved shirt and a pair of old khakis got out from behind the wheel. He was yelling and making his way to me as I exited the driver's side of the truck. I held out my hand to him, to shake hands and calm him down, but he ignored it and continued to shout. Then other people started to gather from a few nearby houses: a tall gaunt man in a cowboy hat; a heavyset woman with a couple of kids dressed in camouflage; some other men and women looking mad and rural. Soon a group of ten to fifteen people were clustered around our trucks, many of them yelling at us and fiercely angry. We were in an isolated area, unarmed, and did not have radio or cell phone contact. I realized I was going to have to talk my way out of this one.

CHAPTER TWO

The Importance of Effective People Skills in Conservation

Overuse and degradation of the world's natural resources is becoming critical. In 1980, Paul Ehrlich predicted that the earth's population would reach six billion by 2000.[1] His predictions were very close. The six-billion mark was reached on or about October 12, 1999, and it continues to climb at an annual rate of 1.4 percent per year.[2] This equals two hundred thousand more people each day. Every forty days, a city the size of New York added to the earth. Every four years the entire current population of the United States is added to the globe. More people were added to the globe during the twentieth century than in all previous human history combined.

All this would be less of a problem if the earth did not have a limited carrying capacity, or total mass of people it can support. People deal with the concept of carrying capacity every day. Your garden has a certain total amount of plants it can support. The only way this "base" amount of plants can be increased is through the addition of fertilizer. If there are too many weeds, the amount of desirable plants that can be grown is much less, because the weeds use the food and water that would ordinarily be available for the desirable plants. When we set up a goldfish bowl as kids, we knew the rule of thumb was that there could be no more than one inch of fish per gallon of water. Otherwise the fish tank would be too crowded for the fish to survive. Every waterbody, or landmass, including the planet Earth, has a set amount of organisms, including humans, it can support before mass die-offs occur. However, it is likely that before the carrying capacity is reached, overcrowding, reductions in quality and quantity of food, and pollution would greatly reduce the quality of life.

At the same time the human population is increasing at a geometric rate, species of other animals and plants are disappearing. Many say that species have always disappeared—this is a common occurrence. The dinosaurs are no longer on the earth, wooly mammoth are found only in museums, and the giant ground sloth can be found on savannas no more. However, it is not that species continue to disappear that is disturbing. It is the increasing *rate* at which these disappearances are occurring that is of concern. Pulitzer Prize–winning ecologist E. O. Wilson cites three different independent methods that conclude the current rate of species extinction is somewhere between one thousand and ten thousand times faster than the extinction rate before the presence of humans.[3]

Other evidence suggests that all should be concerned about degradation of our natural environment. Global warming has been well documented, and if it continues, scientists argue it could lead to the flooding of coastal cities, increases in the frequency and magnitude of storms, and altered precipitation levels, making some areas drier and others wetter. Some people reason that global warming results from natural climate cycles, but a growing number of people point to evidence such as carbon dioxide emissions growing twelvefold between 1900 and 2000[4] and argue that human production of greenhouse gasses is responsible.

HOW HUMAN BEHAVIOR AFFECTS THE ENVIRONMENT

How can we reverse these disturbing trends? The first step is to recognize the factors that contribute to the depletion of our natural resources. The impact (I) of any population on natural resources was defined by Paul Ehrlich and J. P. Holdren[5] as the product of population size (P), its affluence or per-capita consumption (A), and the environmental damage (T) inflicted by the technology used to supply each unit of consumption: I = PAT.

Human behavior influences each of these variables. The reproductive behavior of people dictates the growth of the population (P). The amount of resources used per person (A) is a function of their behavior. The technology used to extract or process the resources (T) is a function of not only what technology is available, but what technology people adopt.

WHY SHOULD WE PROTECT THE ENVIRONMENT?

Currently, how important is the conservation of natural resources to people? Do most people feel that conservation of natural resources is a "luxury," or something required for the very survival of the human species? Most people consider conservation of natural resources important, but in Gallup polls conducted during the past five decades, environmental concerns have never registered as the number one issue for more than 7 percent of the people.[6] In addition, those employed in the natural resources profession have often gotten a bad reputation. I once had a radio talk show host from a religious station ask me what I did for a living. I told him I taught college students about natural resources. He said, "I haven't had a chance to talk with an environmentalist. Why do they consider animals more important than people? Why do they think saving an endangered species is more important than a person's job?" A former secretary of the interior discussed natural resources conservation with me during a telephone call and said, "I'll tell you what's wrong with this country. It's the environmentalists. They're destroying agriculture."

Protecting the environment is not a Republican, Democratic, or any other party issue. It is not a liberal or conservative issue. It is not a rural or urban issue. Protecting the environment is in everybody's interest and is everybody's problem. Economic, scientific, moral, and religious arguments have all been made to support respect and protection for the environment.

Often those who say that environmental regulations should be relaxed state that it is a choice between jobs and the protection of natural resources. However, prominent economists have found this logic faulty. More than one hundred economists across the United States, including two Nobel laureates, wrote a letter to President George W. Bush and the governors of eleven western states in 2003.[7] In this letter they declared that there was little evidence to support the claim that economic declines occur if the decision is made to protect natural resources. In fact, the West's natural environment is, arguably, its greatest, long-term economic strength. According to these economists, economies of western states are strongly dependant on a high quality of life and tourism. Instead of net jobs lost when environmental protection is enforced, jobs lost in a consumptive industry, such as mining, are counterbalanced by jobs gained in ecotourism, and by high-tech companies that move to the area because quality of life is better. According to economists Ernie Neimi, Ed Whitelaw, and Andrew Johnston this was strikingly illustrated by the northern spotted owl controversy in the Pacific Northwest. Despite many predictions that the economy in the Pacific Northwest would experience a huge downturn when logging was restricted to protect old-growth timber and the northern spotted owl, impacts were much less to the region than predicted, and economic growth actually occurred.[8] Presumably, these same arguments could be effectively made for other regions of the globe as well. Ernie Neimi further underscores this point by saying, "When I go into a room and ask people if they would like a 20% increase in income, everybody's hand goes up. When I then say 'How many of you are willing to move to New York City to realize this increase?' the hands quickly drop, with few exceptions." This shows the substantial economic value of the surrounding environment and other amenities to people. It also shows that, through their management of these amenities, communities can attract households (or drive them away)." [9]

The world's major religions all advocate respect for the world's environment.[10] In the Judeo-Christian religious tradition, the Bible states in Genesis that man was placed in the Garden of Eden to take care of it—to be a steward of the Earth's resources. In Islam, man was given a similar stewardship trust over the environment. The Koran advocates that man can use from the environment, but not to excess. Out of concern for the total living environment, Buddhists often extend loving-kindness beyond people and animals to include plants and the earth itself.

Some argue that we can use as many and whatever resources we want—God will take care of us. Others say this argument is a bit like the old joke about the man who climbed the roof of his house when it flooded. Two boats came past and the people in both said, "Jump in and we will save you!" The man said, "No, God will protect me." Another boat came by, and the people in it said, "Jump in the boat, this is your last chance to escape!" The man said, "No, God will save me." Finally, the flood waters came over the roof and drowned the man. When he went to heaven he asked God, "Why didn't you save me?" God said, "What do you mean? I sent three boats!" Many religious leaders argue that God has already given us the needed tools to protect our environment—it is up to us to decide to use them.

For those who are not religious there are plenty of other reasons to protect and conserve the environment. Most people know that on a large scale, the quality of their environment is important for their very survival. However, some can question why it is important to save individual species from destruction. With some species, it is easy to make an argument for their protection because they are loved, charismatic, or important symbols. Can you imagine our national symbol, the bald eagle, going extinct? Not being able to go anywhere to see grizzly bears, tigers, whales, or giant redwoods? However, for less charismatic species, there are strong arguments as well.

Individual species can serve as "canaries in the coal mine," warning of degraded conditions that may not be fit for humans. The presence or absence of certain species of insects in streams is regularly used to indicate stream pollution. Small plants called lichens are being used in an attempt to detect high levels of metals in the air around southern Arizona towns.[11] These metals are thought to be responsible for increased death rates in the area due to childhood leukemia. It is hoped that the lichens will be able to absorb metals more effectively than conventional detectors.

Individual, seemingly insignificant species can have important, yet undiscovered uses. Taxol, an important drug in the fight against cancer, is found in the diminutive, unimpressive-looking Western yew. The yew was once a "weed tree," cleared as trash before the discovery of the importance of this chemical found in its bark. In 1992, Taxol finally hit the market and was trumpeted as the most important cancer-fighting drug in two decades. The tree was then important for saving lives and was treated with new respect. Demand for the tree became so great, and supplies so limited, that an artificial method was developed to synthesize the compound found in the bark.[12] If not for the yew, the existence of this chemical would have never been known.

When an individual species disappears, the effects can ripple throughout the entire ecosystem, affecting many other plants and animals. Some biologists assert that this is similar to the effect removing one small part of a car, such as the spark plug, would have on the entire vehicle.

So after all this, what if the species seems so insignificant that it has no human uses, contributes little to the ecosystem (which incidentally is not very likely), and is rarely seen? Should we not be concerned about its destruction? Even the smallest species are fabulous genetic masterpieces. The ptiliid beetle, which is smaller than a period on this page, has six legs, a pair of wings, fully functioning reproductive and digestive systems, and a genetic code that when printed in letters of standard size would stretch 1,600 kilometers.[13] Some have compared these genetic masterpieces to art in the Louvre. Most of us have not seen the wonders of the Louvre, such as the *Mona Lisa*, the *Venus de Milo*, and the *Victory of Samothrace*. These paintings and sculptures have no apparent everyday "use" to us. Even though few of us have seen them or few use them, there are few people who would advocate bulldozing the Louvre to make way for a parking lot.

HOW GOOD ARE NATURAL RESOURCES PROFESSIONALS AT MODIFYING HUMAN BEHAVIOR?

Although convincing arguments can be made for protecting our environment, the key is getting people to listen to these arguments and take action. The behavior and attitudes of people are incredibly important for protecting air, water, species, and, in fact, our very existence. Because of the importance of people's attitudes and philosophies in conserving our natural environment, one would have thought that conservation professionals would be the best psychologists, communicators, marketers, lobbyists, and negotiators of any profession. While there are very talented conservationists working in these areas, skills related to working with and influencing people are often less developed in the natural resources profession than in other occupations. I surveyed the natural resources programs of thirty-one fish and wildlife studies programs at twenty universities. Most programs required about 120 credits to graduate. Beyond the general requirements of the university, no psychology or marketing classes were required in any of the programs. Human dimensions, policy, and administration courses made up no more than three to six of the required hours. Psychology or humanistic skills were usually required by the universities in their general requirements for all majors, but these requirements could be met just as easily by taking courses such as Medieval Philosophy instead of Social Psychology and Public Relations.

When I graduated with my doctorate from the University of Washington, I thought I knew a lot. The University of Washington had one of the finest and oldest fisheries schools in the nation. I had a bit of a cocky attitude, and on occasion could be rather condescending. My first job after I finished was with the Washington State Department of Wildlife, an agency charged with managing the wildlife and the inland fishes of the state of Washington, while at the same time keeping happy millions of customers who enjoyed hunting and fishing each year. Once I set up my desk in my new cubicle, things changed in a big way from the somewhat protected climate of academia. I had to manage a staff, deal with angry constituents, present information to state senators who had immediate deadlines, and write rapid, factual reports. I realized that although I received an excellent training in science, I did not learn in school everything I needed to succeed in the hard, fast world of government bureaucracy. Realizing I had a pretty good science background under my belt, I set out to learn as much as I could about how to collaborate, negotiate, and communicate effectively: skills that ordinary government workers and employees such as fish and wildlife biologists, foresters, geologists, teachers, firefighters, procurement officers, and other mid-level bureaucrats had to employ every day to be effective.

I soon noticed that there were some people in our agency who were extremely effective and efficient. These were not necessarily the most technically competent, or the flashiest, although they usually had good skills in one of these areas. What I noticed about these people is that they were best in dealing with people. Their people skills propelled them into positions of authority, and enabled them to make a difference where others could not.

When most people get into the conservation profession, they say they were attracted to the field by opportunities to work outdoors with animals, plants, soils, or rocks. Few say they get into the conservation profession because their greatest love is working with people; however, those employed in conservation jobs deal continually with people and their behavior. Agency biologists work with landowners to improve their property for wildlife. Timber biologists negotiate cutting schedules and areas with representatives from other groups and agencies. Agency staff conduct workshops, hunter safety courses, fishing clinics, and other types of training for citizens. The ability to work effectively with people is as important for the conservation professional as it is for the policeman, the school teacher, or the lawyer. In fact, a well-developed ability to work with people is the only way that conservation professionals can be effective in carrying out their job.

Still unconvinced? When I read historical accounts and spoke with various natural resources managers, I discovered that many spectacular successes and failures in conservation could be attributed to successes and failures in working with people. Perhaps this could be illustrated by some examples. In both of these examples, talented, intelligent biologists worked extremely hard to solve important management problems. In one example, the politics of the situation happened to work out. In the other it did not.

EXAMPLES OF PEOPLE SKILLS IN ACTION: LAKE DAVIS AND DEVIL'S HOLE

Lake Davis, California, is a seven-mile-long reservoir high in the Sierra Nevada Mountain Range in eastern California.[14] It is located in a beautiful area covered with pine forest and provided 50 percent of the water supply for the small town of Portola, located seven miles from the lake. Portola was named after the discoverer of San Francisco Bay and the first governor of California, Gaspar de Portolà. Historically, the town served as a stage stop for gold rush travelers and later provided a rail yard for the former Western Pacific Railroad. Nearby reservoirs and lakes, such as Frenchman Lake and Lake Davis, were known for their outstanding trout fisheries. In 1988, northern pike were discovered in Frenchman Lake.[15] Northern pike are in many parts of the United States, but before their entry into Frenchman Lake they were not known to have been in California. The discovery of northern pike in Frenchman Lake was disturbing. The northern pike is a voracious predator with a duckbill snout containing numerous sharp teeth. The fish can grow to lengths over three feet. Northern pike introduced into lakes in other states had quickly decimated other fish populations.[16] The danger of having northern pike in Frenchman Lake was that the reservoir drained into the Sacramento River and Sacramento Delta.[17] The Sacramento River, in addition to most of northern California's coastal rivers, contained imperiled runs of Pacific salmon, steelhead, and other rare native fishes. If the Northern pike escaped from the reservoir, rare fishes in California's coastal rivers could be threatened.

As the conservation agency responsible for fish and wildlife in the state, the California Department of Fish and Game (CDFG) had the task of deciding what to do with the northern pike. The CDFG proposed to treat the lake with piscicide (fish poison) to remove the pike, and filed an Environmental Impact Report to identify the potential effects of the treatment. The use of the piscicide was challenged and treatment was held up until 1991 when the lake was treated.[18] By then, northern pike had already started to spread. In September 1991, an angler caught a pike in the Middle Fork Feather River, which was fed by water from Frenchman Lake. A small treatment was conducted, but further CDFG sampling confirmed the presence of more pike in the river.[19] The department treated the Middle Fork Feather River and Sierra Valley waterways in 1992, and reported that pike were successfully eradicated.[20] The department had found an effective tool to keep the pike at bay and expected that if the pike showed up again, they could be handled effectively with minimal problems. However, as events turned out, they could not have been more wrong.

In 1994, northern pike were captured in Lake Davis. The CDFG discussed options and decided again to use piscicide to control the pike. Nusyn-Noxfish was an effective fish-killer. It contained powdered rotenone, which was made from the root of the Derris tree, a plant used by Native Americans and South Americans for centuries to capture fish for food. It also contained a synergist to help it work better. The synergist contained trichloroethylene, which was a possible cancer causing agent. However, according to state biologists, the amount of trichloroethylene that would be put in the lake would be one-tenth of that allowed in municipal drinking water by the Environmental Protection Agency (EPA).[21] Other treatments such as mechanical removal were not nearly as effective, and the rotenone and synergist would rapidly dissipate following treatment. It sounded as if an effective solution had been found.

Local residents were not convinced that Lake Davis should be treated using rotenone to remove the northern pike.[22] They formed the Save Lake Davis Committee, first called "Victims of Lake Davis," in early 1995 to prevent the rotenone treatment.[23] Shortly after, officials from Plumas County and the city of Portola joined the group and it was subsequently renamed the Save Lake Davis Coalition. Their argument was that Lake Davis supplied the county with drinking water; they did not want carcinogens in their water, no matter how low the concentration. Members of the public also felt that rotenoning the lake would have a substantial economic effect on local recreational businesses. They also argued that although the state biologists would not admit it, northern pike were already in the Sacramento River delta; pike in the lake did not eat trout in Lake Davis but ate insects; and many western states co-managed trout and northern pike.[24]

During the spring and summer of 1997, local residents and officials stepped up pressure on the state. In March 1997, Plumas County and the city of Portola filed suit to stop the project.[25] A letter writing campaign to legislators was begun, noisy marches on the state capitol ensued, and even a song was written by a local resident to protest the treatment. It went:

Hey Mr. Fish and Game
What you want to do is really lame
Would you take a cup of poison water?
Would you give it to your son or daughter? [26]

Local officials passed an ordinance in September making it illegal to treat the lake. They vowed to have Plumas County sheriff's deputies arrest any California Fish and Game staff who dared to try to treat the lake.[27] However, the protests and suits were in vain. The state upheld the right to treat the lake, obtained the necessary permits, and countered that it would use the Highway Patrol to arrest any Plumas County sheriff's deputies that interfered. It promised to provide alternate water supplies and other mitigation. A day before the treatment, a local judge ordered the county not to interfere with the state or its officers.[28]

On October 14 the treatment began. State personnel transported several truckloads of piscicide to state property near the Lake Davis Dam at 2 a.m. to avoid confrontations with the protestors.[29] Because of the volatility of the situation, the one hundred Fish and Game staff brought to the lake to apply the toxicant were protected by ninety game wardens, fifty-five Highway Patrol officers, and twenty-five sheriff's deputies.

Hundreds of protestors gathered on the icy shores of Lake Davis before dawn on October 15, 1997, for a candlelight vigil.[30] The protestors chanted and jeered at Fish and Game officials and pelted them with Halloween candy. As the protestors cheered, four wetsuited demonstrators, including a Portola city councilman, swam out to the middle of the lake and padlocked themselves to a buoy in a final effort to prevent the treatment. They were arrested for trespassing, and three of them were treated for hypothermia at a local hospital. Three other protestors were arrested for moving markers on the shore that the boats needed to use in the operation.[31] By 8 a.m. twenty-two Fish and Game boats motored onto the lake and began applying the Nusyn-Noxfish to the water. Following applications of the chemical, dead fish were removed for three days. The majority of fish killed were northern pike.

Agency staff had promised that all chemicals involved in the treatment would dissipate within a month. However, trace amounts were being found through July of the following year. Plumas County officials filed criminal charges against the state alleging negligence, and the state finally settled with a $9.1 million payout.[32]

In May 1999, seventeen months after treatment, northern pike were again found in Lake Davis. Several hundred northern pike were captured in the summer of 1999. To this day, piscicide treatments of western lakes have been much more difficult to conduct because of public perceptions of piscicide following the Lake Davis incident. Stories were circulated that state biologists could not get their trucks filled with gasoline by service stations in the town, and had to go to other areas. Now California Fish and Game and the townspeople are working together more successfully, but tensions are still present in the area,[33] and the northern pike are still in the lake.

Less than five hundred miles away from Lake Davis, another fish story worked out magnificently for the protection of a species. However, by all accounts it should not have. The subject was a fish no one liked to eat and few people saw, which lived in an area not known for protection of fishes. By the end of this story, however, this lowly rare fish, less than one-inch long, made it all the way to the U.S. Supreme Court.

If you are driving east from Death Valley Junction, California, in the distance, across the Nevada line, you see a few splashes of green brush that stand out in the vast stark beautiful desert of creosote bush, multi-hued sand and rocks, and brilliant blue sky. When you drive closer to the green, you see you have arrived at Ash Meadows. From the surface, Ash Meadows is unimpressive. It is not majestic like the Grand Tetons, or awe inspiring like the Grand Canyon. There are no giant redwoods or massive Sitka spruce.

However, this small damp area, in the middle of the hottest, driest land in North America, is one of the biological hotspots of the country.

The treasure of Ash Meadows is subtle, only for the discerning eye. In this small oasis, over twenty-three endemic plants and animals have evolved, the largest such assemblage in so small an area in the United States.[34] In the Pleistocene epoch, large lakes covered the region. When the lakes started to dry, aquatic animals and plants were pushed into smaller and smaller pools and springs.[35] The organisms that are left today are survivors from those large lakes, and have evolved to live in their new harsh conditions. They are castaways on small aquatic islands in a vast burning desert. One of the most interesting aquatic islands, located on the far northeast corner of Ash Meadows, is Devil's Hole.

Devil's Hole lies in a narrow seven-yard-long crack of rock, a small window to an immense underground cave system. Anyone would expect any water sources to be known in a desert such as this from the earliest times, and such is the case of Devil's Hole. From the days of the gold rush miners traveling west, the hole provided a welcome resting spot and was first called Miner's Bathtub and then Devil's Hole. Because of its unique character, it was designated as part of Death Valley National Monument in 1952.[36]

One of the reasons Devil's Hole is so interesting and unique is because of the Devils Hole pupfish that live in the system. This pupfish species is found nowhere else in the world except in this pit. While the cave system is vast, the opening to the surface is not. And what is located at the opening limits the pupfish population and is critical for their survival; most of the algae that feeds the fish, and the spawning areas for the fish, are on a ten-square-yard shelf of limestone at the top of the hole.

Water in the hole is old—fossil water some call it. This water is not from recent rains. It is ancient, originating from rains ten thousand years earlier, percolating and creeping through the mountain rock until finally reaching the hole and the adjacent springs at Ash Meadows. Any water in the desert is coveted by those who wish to profit. And many of these individuals do not care if it is from yesterday's thunderstorm, or if it fell ten thousand years previously.

Until the late 1960s, water levels in the hole remained fairly constant. But soon afterward, various biologists and naturalists grew concerned. The water level in Devil's Hole had started to decline. Once it fell to the limestone shelf, it was likely that the Devils Hole pupfish were finished.

What was causing the declining water levels? The philosophy of many at that time (and of some people today) was that water unused is water "wasted." Entrepreneurs knew of the springs at Ash Meadows and Devil's Hole and thought there was money to be made by growing alfalfa. To irrigate their crop, they started pumping out to their fields the water that was just sitting in the springs "being wasted." One pump was even built directly over the top of Jackrabbit Springs, a nearby water source in Ash Meadows. Most of the pumping was conducted by a company called Spring Meadows.

As the water level in Devil's Hole continued to decline, biologists knew they had to act quickly. At the time, budgets of agencies were small, and water laws were weak, especially when used to protect small native fishes with little or no commercial value. However, there was universal agreement among agency biologists, university researchers, and the general environmental community that the Devils Hole pupfish and other aquatic species in the Ash Meadows area were too valuable to be lost. A Pupfish Task Force was formed by the Department of the Interior, and a special council, devoted to saving fishes of the desert was formed. The new Desert Fishes Council consisted of agency representatives, academics, and private biologists, all dedicated to the goal of saving populations of endangered desert fishes.[37]

The Desert Fishes Council, as well as other environmental groups devoted to saving the unique fauna of Ash Meadows, knew they had to publicize the plight of the fish to develop public support. Just as in the Lake Davis scenario, there was considerable local opposition to the stance the biologists and agencies were taking. It ran counter to free enterprise. "Kill the Pupfish" bumper stickers appeared on the streets of local towns. One local regional manager of a land management agency said he could not believe all of the fuss over a bunch of worthless fish. An extreme position was taken by the editor of the *Elko Nevada Daily Free Press* that suggested the controversy could be done away with if someone dumped rotenone in Devil's Hole.[38] However, press releases from the task force and state fish and game agencies and the unusual, rare nature of "fish in the desert" started to capture the attention of the national news media, and a groundswell of public support for saving the pupfish began. There were articles in major papers such as the *Wall Street Journal* and the *Los Angeles Times* about the controversy. *Cry California* published a series of articles in the spring 1970 issue that discussed what was happening to the fish. An NBC documentary in

1970 titled *Timetable for Disaster* discussed the plight of the pupfish on national TV; later that year it was given an Emmy Award for outstanding documentary of the year. In a meeting with the secretary of the interior, Walter Hickel, the show's producer-director, Don Widener, said that he would do another documentary if the Devils Hole pupfish was allowed to go extinct. However, Widener said this next documentary would focus on examining why the federal agencies allowed this to happen.[39]

Letters supporting pupfish protection flooded into the offices of state and national politicians. As the public influenced politicians to protect the pupfish, work inside the executive branch and the court system continued as well. Fortunately for the Devils Hole pupfish, members of the Desert Fishes Council were on a first-name basis with some of the leadership in the Department of Interior. Charles Meacham, who was commissioner of the U.S. Fish and Wildlife Service at the time, even played around the spring systems near the California-Nevada border as a child and knew Devil's Hole and Ash Meadows were too valuable to lose. As soon as he found out about the problem he organized the above-mentioned task force to do something about it.[40] U.S. attorneys argued in the courts that Spring Meadows Inc. could not take water that would affect the purpose of the Devil's Hole reservation. An injunction was secured to stop the drilling, and Spring Meadows (later changed to Cappaert Enterprises) appealed the injunction all the way up through the U.S. Supreme Court.[41]

Phil Pister, local California Fish and Game biologist, and executive secretary of the Desert Fishes Council, later gave an example of how judges and politicians were successfully influenced by advocates for the Devils Hole pupfish. Pupfish advocates identified the interests of the decision makers and found ways to argue the case in those terms. One of the judges hearing the case was deeply religious. Therefore instead of arguing the merits of biodiversity and the uniqueness of a fish evolving for eons out in the desert, Pister made arguments that would appeal to the interests of the judge. Don't we have a stewardship obligation to protect God's creatures, which is outlined in Genesis? Wouldn't God be upset if we did not meet our obligations? Following these sorts of arguments, the judge ruled in favor of the pupfish.

In 1976, the U.S. Supreme Court blocked the final legal challenge by Cappaert Enterprises.[42] Following some years of wrangling for land ownership, the U.S. Fish and Wildlife Service acquired Ash Meadows and turned it into a wildlife refuge. How biologists, managers, and conservationists worked together to help politicians and the American public realize that the Devils Hole pupfish and Ash Meadows was worth saving is a valuable lesson. Today, the strategies employed to save this unique area and fish species are regarded as a hallmark in environmental protection.

Why did one attempt at a conservation activity (Lake Davis) have such problems, while the other (Devil's Hole) worked out well? The repercussions of each case history had implications across the United States and for years to come, and influenced the fates of other conservation activities. The success at Devil's Hole has translated into protection for dozens of other rare plants and animals. The political setbacks at Lake Davis made it more difficult to use rotenone across the western United States.

CONCLUSION

In this book we will explore political, communication, and negotiation techniques that have been used to protect and conserve our natural resources. Some of these skills for working with people were gleaned from successful natural resource programs, and others from events throughout history. To compile these techniques, I carefully studied the practices of conservation professionals, leaders, and government workers who were most effective. I examined their careers with the goal of not just determining what they did but *how* they accomplished it.

In addition, I will show you techniques for working with people that were developed in other disciplines such as psychology, sales and customer service, communications, marketing, and political science. We will explore how you can use these techniques to increase your effectiveness and job satisfaction and develop friendships with a wide variety of people.

I have used all of the skills presented in this book at one time or another, and have been very satisfied with the results. You are faced with the challenge of getting others to protect and conserve our priceless environment for future generations. Hopefully you will find the skills presented in this book will help you meet this important goal.

CHAPTER SUMMARY

- Paul Ehrlich and J. P. Holdren identified the most important factors affecting the degree of impact on our natural resources. Every factor in their equation, impact = human population size x impact or use per person x impact of technology used to extract the natural resources, is influenced by human behavior. Being able to influence humans to behave in an environmentally friendly manner is therefore necessary to protect our natural resources.

- Preserving natural resources is necessary for our survival, and not just a "luxury." Numerous prominent economists have found a healthy environment necessary for economic prosperity. All the world's major religions support protection of the environment. Individual species can hold cures for diseases, indicate declining environmental quality for humans, and fill important roles in functioning ecosystems.

- Because human behavior is such an important factor governing impact on our natural resources, learning how to work well with people and influence them is critical for the natural resources professional to make a difference in the conservation arena.

- Natural resources efforts can succeed or fail, often depending on how well the conservation professionals work with the stakeholders and others. The success at protecting water levels in Devil's Hole, Nevada, translated into protection for dozens of other rare plants and animals. Inability to arrive at a jointly acceptable solution among agencies and locals to remove northern pike in Lake Davis, California, resulted in a multimillion-dollar legal settlement and made it considerably more difficult to use piscicides across the western United States.

CHAPTER THREE

How to Resolve Conflict and Defuse Contentious Situations: Verbal Judo and Other Communication Techniques

When I got out of the truck, I could see the man was extremely angry. His face was flushed, his eyes were wide open, and he approached me yelling insults and threats. I reached out my hand to shake his, but he refused it and continued yelling. The other biologists were in the truck, and people from nearby houses started to gather. They all looked angry and menacing. They were yelling that whatever the government, that is, "big brother," wanted to do to "their" lake was not okay with them, and they wanted us to get out of there right now. We were out of cell phone contact, out of radio contact, and it was important that we work on their lake, because it had the longest-term data set on outmigrating salmon of any small lake in western Washington—a critical component of our study.

The man left his pickup truck and came up to me screaming, "What the hell are you government people doing here? You have no right to bother us! Get the hell out of here! This is a private lake!" At that moment I had three choices: (1) I could yell at the man and the crowd and hope my anger would drive them back; (2) I could get back into the truck and get my crew out of there as quickly as possible; or (3) I could use "verbal judo" to try to smooth out the situation.

What if I had tried anger or the "mad" approach? Here is a likely scenario. I could have said, "Don't you threaten me! What the hell do you think you're doing! As an agency biologist, I've got the perfect right to sample fish in the lake. The law's on my side! Get out of my way, or I'll call the game wardens and get you arrested!" This response may have made me feel considerably better in the short term, because I could vent on the guy and I would not have let him "get away" with insulting me. I may have even made him back off a bit, because he might have been concerned that I would make good on my threat. However, this would have probably poisoned my relationship with him in the long term. He may have come back at me even angrier. At the very least, even if he backed off at that time, he would have probably felt justified about harassing us over the two-year period we had to work at the lake, maybe with words, maybe by vandalizing my truck. While some in the crowd were angry along with him, others were curious and just wondering what would happen. If I, a government employee, came back at him with extreme anger, it is likely that some, if not all, of the crowd would have felt compelled to defend him. If we would have made good on our threat to call in wildlife agents to get us out of the crowd, we probably would have been humiliated at work by not being able to handle the situation ourselves. We would have also damaged hope of a good relationship between those people and State Fish and Wildlife in the future. Furthermore, agency management would most likely have made us choose a different study site in order to avoid controversy in the future. The anger option could go very wrong, with little chance for success.

The other option I had was to back off, or the "sad" approach. As soon as the man approached yelling, I could have walked hurriedly back to the truck and said, "Let's get out of here!" We could have wheeled the trucks around quickly and avoided the wrath of the crowd. This could have given us the highest margin for safety. We would have avoided further verbal conflict that could have escalated. However, this would have been the most damaging choice for my self-esteem, as well as damaging the respect my crew had for me. It is unlikely we would have continued to use the lake, so we would have lost one of the best study sites for the project in western Washington. The crowd would have lost respect for government biologists, and in future encounters would have learned that meeting biologists with anger and disrespect would be the best way they could get what they wanted. Stories travel quickly among agency biologists, so my social standing in the agency could have also been affected.

Clearly, there are times both the "mad" and "sad" approaches work. If I needed to bluff my way out of a situation that I would not need to come back to, and the body language of the critic told me I could get away with it, a strong counterattack might have been the best approach. If we were accosted by camouflage-fatigued sociopaths carrying assault weapons, the flight approach would have been the option of choice. But for most situations, including this one, there is a third choice—verbal judo. I have found verbal judo (the "glad" technique) to be most effective in the majority of cases.

VERBAL JUDO

What is verbal judo? Verbal judo is using words to calm down and "disarm" your critic. The technique is analogous to a martial art in that you disarm your critic without running away or getting angry, and you work with your critic to use his or her own strength to solve the problem. Numerous excellent books and articles discuss the procedure.[1] While all have similar steps, I borrow most heavily from the process used by psychologist David Burns.[2]

Understand Your Critic

Before you can help your critic or develop a better relationship with him or her, you need to know from where he or she is coming. This can be done in two steps: ask questions to understand why they are criticizing you, and then demonstrate to them that you understand.

Ask Your Critic Questions to Draw Out the Real Problem
If the person is explaining the problem or venting, you actively listen, asking questions when appropriate to draw him or her out. In his book *Seven Habits of Highly Effective People*,[3] Steven Covey says to listen so you can be understood. This period of questioning and listening allows the other person to vent their anger and allows you to hear their point of view so you can effectively respond. How often have we jumped in to solve a person's problem,

only to find out what we were trying to solve was not the thing they were upset about in the first place? Phrases to draw someone out might include:

Wow, you sound upset. Is something bothering you, and if so, what is it?

I appreciate you being candid with me. Please feel free to keep going.

Can you tell me a bit more?

I had a supervisor at the Washington Department of Fish and Wildlife who used this technique masterfully. At Inland Fish Division meetings, the manager would ask agency biologists their opinions on different matters. All would have a chance to vent and blow off steam, and once this was over, problem solving could begin. I never saw the supervisor lose his temper, even though criticism of management staff and blunt comments were frequently aired by district and regional biologists.

David Burns, in his book *Feeling Good*,[4] suggests that specific questions are most important in getting to the root of the problem. For example, someone might say, "You Fish and Game biologists don't know your butt from a hole in the ground." This is a nebulous criticism that really doesn't tell you anything. You want to ask them specific questions to identify the underlying problem. Therefore, you might say, "Wow, you sound pretty unhappy with Fish and Wildlife biologists. What happened to make you so angry at us?" After this he may respond, "You jerks closed hunting in this area." Now you have a specific concern to address. Here are other examples of what you might say to get specific.

Criticism: "Jones, you don't seem to get anything done."

Your question: "I can see you are upset, Mr. Crabbus. What did I do that led you to that conclusion?"

Critic: "You did not turn in the Fiplet Lake Water Quality Report. It was due yesterday." (Here is the underlying problem—now you are able to work to solve this problem.)

Sometimes it takes a couple of questions to draw out the real issue. Keep asking questions until you have something specific with which to work.

Criticism: "The research unit is a waste of resources and never does anything worthwhile."

Your question: "Wow, it seems like you don't have much respect for the research unit! Why do you feel that we don't do anything worthwhile?"

Critic: "You guys seem pretty cocky and you don't work well with agency geologists." (Here, the geologist shows that they have had their feelings hurt in the past, and you can attack that specific problem.)

Your question. "Thanks for letting me know. Is there something on which you think we didn't work well with you in the past?"

Critic: "Yeah, when you were setting up that meeting to discuss soil management, and you know I am one of the experts in the agency on soils, and I wasn't invited. Why not?" (Now you have something with which you can work. More than likely the geologist has had his feelings hurt because he was not invited.)

It is critical that you actively listen and keep your cool during this period. If you can do this and effectively use the steps below, you can often turn a critic into a supporter.

Paraphrase What They Say to Demonstrate That You Understand Their Concerns

During the understanding stage, paraphrasing, or repeating back to the critic what they said in your own words, is wonderfully effective.

Critic: "I work and work all day, and the government just wants to bury me in paperwork. How is a landowner expected to keep up with all of this? What a worthless bunch the State Wildlife Agency is!"

Paraphrase: "What you're saying is that the government paperwork is way too time consuming to fill out, and it is really hard to keep up with all of it—is that right?"

Notice that this is not repeating word for word what the person says to you, which is parroting, and sounds artificial, but is repeating back to them what they said in *your own words*. Here are some more examples:

Example: "I don't seem to get any recognition from you for what I do. You promised me a raise, and you never seem to deliver."

Paraphrase: "It sounds like you're pretty frustrated because you feel that I don't recognize your good work, and you don't think I follow through on things, especially your raise. Correct?"

By asking at the end of the paraphrase if you are correct, you give the person the opportunity to correct any errors you might have made in paraphrasing his or her concerns.

Sometimes you don't need to ask questions because you are perfectly aware of what the other person is upset about. If this is the case, one option is to simply skip to the next stage.

The Verbal Judo Stage or Disarming Your Critic

Once you know the true nature of the problem, and the person has had a chance to vent if needed, it is time to advance to the disarming stage. There are three extremely effective techniques for disarming your critic and getting his or her to your side. Any one of them can be used individually, or they can be used in combination.

Agreeing in Some Form with What They Say

I knew an older professor who was a court advisor to Judge Boldt in Washington State. In the mid-1970s Judge Boldt ruled on probably the most significant court decision affecting salmon fisheries in the state of Washington. He had the task of interpreting Indian treaties made one hundred years previously to determine who had the right to catch certain segments of the salmon population. Before the mid-1970s the tribes had no special rights over and above those of anybody else. However, Judge Boldt ruled that the Medicine Creek Treaty of 1854 stated that the treaty-signing tribes were entitled to catch 50 percent of the salmon. After the decision, the tribes were allowed to fish for and keep about 50 percent of the salmon catch. This angered the Anglo gill netters who were used to catching and selling most of the fish, and had been doing so for generations. The ensuing controversy surrounding this decision was phenomenal, and my friend told me of many heated arguments. He was a mild-mannered individual who generally got along with people very well. I asked him how he could remain so popular and retain his composure in the face of so much stress and controversy. He said basically that he got along because if someone called him an SOB, he would find a way to agree with them.

Agreement stops a critic in their tracks. They have nowhere to go—the argument is over if you find a way to agree with them! Many of you might say, "No way, the critic's logic is so far off, his points so nutty, that there is no way I could agree." However, below are some examples that show a person, if creative, can agree with almost anything in some form without lying! This is a crucial step because the technique does not work unless what you say back to the critic is truthful. You will see in the examples below that no matter how outlandish the statement, I can always find a way to agree without lying.

Criticism: "Fish and game people are incompetent asses."

Agreement: "You're right; some of them are!"

Criticism: "These hunting regulations are bulls__t."

Agreement: "Many of them could certainly be improved."

Criticism: "All of you ivory tower professors need to start living in the real world."

Agreement: "You're right! I often think I would benefit from additional experience in the trenches!"

Criticism: "Scott, I felt you did a poor job on this report."

Agreement: "There are several parts of it that could probably be improved."

Criticism: "The endangered species act should be repealed."

Agreement: "There are certainly parts that could be improved."

Now let us try some incredibly outlandish statements to show you that you could take this as far as you want.

Criticism: "The sun is made out of snow."

Agreement: "You're correct in that there is a lot we do not know about the makeup of the sun."

Criticism: "Government biologists are just 'New World Order' people with their black helicopters trying to control the people."

Agreement: "I agree with you that sometimes it is incredibly frustrating working with the government."

Some of you might say, "I will not agree with someone when they are talking about nonsense and insulting me." This is certainly your choice. You are correct that you do not have to put up with someone who is abusive, condescending, or just plain wrong. You have the power to decide if you want to use this tool or not. However, using a tool like this can often allow you to work past the areas of conflict and get you to a state where there is mutual problem solving.

Empathy

The second technique for disarming the critic, and equally important as agreement, is empathy. Empathy is not necessarily agreement, but it is demonstrating that you understand where the other person is coming from or you put yourself in the other person's shoes. You can decide to use empathy with or without agreement. Most people just want to know that you understand them. The following examples show how empathy is used in response to criticism.

The Importance of Truth in Verbal Judo

To some people, verbal judo techniques might seem manipulative. However, you are trying to improve your relationships and ability to communicate with others using these powerful tools. Therefore, it is of utmost importance to be *truthful and honest* in all of your communications. A person lying in agreement statements, slathering on false flattery, or bending the truth to get their way might enjoy some short-term successes, but long-term communication with the parties and your credibility will be damaged. If you are not honest, these techniques will backfire. Approaching all techniques in this chapter and in this book with the sincere intent of improving communication between you and the other party in an honest, direct, straightforward manner will provide the greatest benefits.

Criticism: "Government biologists just want to take people's land."

Empathy: "I know just where you are coming from. My uncle (brother, cousin) was a farmer and he was often frustrated by government regulations."

Criticism: "You guys in the headquarters office don't seem to have any appreciation for what we deal with in the regional offices."

Empathy: "I know exactly how you feel. I worked one summer in a regional office for Smith Office Supplies. The guys at headquarters often didn't seem to understand what we did out there."

Criticism: "I want to camp here for the night. Why are you busting my butt? I'm not hurting anything!"

Empathy: "Hey, I know what you're talking about. It's a bummer hiking all day, pitching your tent, and then being asked to move."

Criticism: "My class has waited six months for you to come over and give us a presentation on water conservation. I'm getting pretty fed up with the delay!"

Empathy: "I hear you. It is so frustrating when you want to plan something like that and it doesn't work out rapidly."

Sometimes it is difficult to agree with a critic, and in these instances empathy alone is an effective technique. A media specialist from the Arizona Game and Fish Department warned me that agreement, used at a public meeting where the TV cameras were rolling, might have an unintended effect. In the interest of having a "sound bite" the cameras might just capture your agreement statement and not capture your actual point of view after the statement. An example might be the following:

Criticism: "I understand the Arizona Game and Fish Department is stocking jaguars in southern Arizona. Why in the heck would they do something like that?"

See how empathy might work better than agreement in this case:

Agreement: "The Arizona Game and Fish Department does stock a lot of things, mainly fishes. I want to make it abundantly clear that they have not stocked jaguars."

Empathy: "I know how frustrating it can be to keep up with all of the activities of a government agency. I want to make it abundantly clear that they have not stocked jaguars."

If the cameras catch only "The Arizona Game and Fish Department does stock a lot of things" and leave out the rest, it could leave the impression on the news that jaguars were stocked. Another caution is when someone wants you to say something bad about a third party. Here is another case where empathy might work better than agreement.

Critic: "Tribal gill netters just want to take half the salmon and waste it. They won't put it to good use like we (the Anglo) gill netters will."

Agreement: "Some tribal gill netters might waste some of the salmon. The judge interpreted that the treaty clearly states that treaty tribes are entitled to 50 percent of the catch."

Empathy: "I know exactly how frustrating it must feel to loose a major part of the salmon catch. My parents were small family farmers, and lost a major part of their livelihood when large agribusiness expanded. The judge interpreted that the treaty clearly states that treaty tribes are entitled to 50 percent of the catch."

If the tribal gill netters or the media only listen to the agreement statement without your later explanation, you could have difficulties. Here the empathy statement might work better and keep you out of trouble.

Don't Criticize, Condemn, or Complain—Use Strokes!

The compliment is a powerful tool for calming down a critic. Using this technique, you compliment the person criticizing you. This can work wonders for calming the situation down and earning an ally instead of an enemy. Here's how it works:

Critic: "You people at Andrews Consulting are just a bunch of biostitutes! You don't care about the environment, but you just want to make money!"

Compliment: "Thanks for your comments. First, I want to say I am impressed with your level of concern for the environment. Most people wouldn't take the time to come out on a rainy evening like this and express their viewpoint. You are to be commended for your interest and commitment! The reason we took this project is. . . "

Critic: "Bass angling has always been an important economic driver of our economy. Why in the heck are you guys screwing with the regulations?"

Compliment: "Mr. Jacobsen, you certainly have a long-term knowledge base about this region and care deeply about environment. The reason we changed the regulations is. . . "

You do not have to invent false compliments when using this technique. *Everyone* does something well. I believe it is better to work hard to identify the person's strengths and compliment those, rather than to make up something that is not true.

Abraham Lincoln— Master of the Compliment

Those of us in biology, conservation, or any area of government can often learn from legendary politicians. A master of the compliment or stroking technique was Abraham Lincoln. When someone told Lincoln that his secretary of war, Edmund Stanton, said he was a damned fool, Lincoln said, "If Stanton said I was a damned fool, then I must be one. For he is nearly always right, and generally says what he means. I will step over and see him."[5]

In another instance, Lincoln was to appoint Civil War General Joseph Hooker as head of the Army of the Potomac. Before his appointment, Lincoln heard that the general said both the army and the government needed a dictator. Lincoln could have beat him back down immediately and refused to nominate him. Instead, he wrote him a letter filled with not only compliments, but also advice. He stated "I believe you to be a brave and a skillful soldier, which, of course, I like. I also believe you do not mix politics with your profession, in which you are right. You have confidence in yourself, which is a valuable, if not an indispensable quality. You are ambitious, which, within reasonable bounds, does good rather than harm. But I think that during Gen. Burnside's command of the Army, you have taken counsel of your

(continued)

Abraham Lincoln
(continued)

ambition, and thwarted him as much as you could, in which you did a great wrong to the country, and to a most meritorious and honorable brother officer. I have heard, in such way as to believe it, of your recently saying that both the Army and the Government needed a dictator. Of course it was not *for* this, but in spite of it, that I have given you the command. Only those generals who gain successes, can set up dictators. What I now ask of you is military success, and I will risk the dictatorship." Lincoln made his point and the letter reportedly touched Hooker with its fatherly tone.[6] Hooker was later placed in charge of some of the president's funerals, ending with the final procession in Springfield, Illinois. Stanton, although he condemned Lincoln initially, later became a strong supporter. In both these instances, Lincoln kept his cool and his respect, and worked through the problem with stroking to make his point and get an end result with which he was comfortable.

Empathy, agreement, and compliments are some of the main reasons that this technique is called verbal judo. By using empathy, agreement, and compliments with people, you are able to immediately get them on your side, in a mutual problem-solving mode, instead of engaging them in a frustrating match of butting heads and trying to prove one's own point.

Diplomatically State Your Point of View

Understanding the other person, and agreeing, empathizing, or complimenting the person builds the foundation or "sets the stage" for you to share your point of view. The person should now be more prepared to listen to you. Now in a kind, matter-of-fact way you can state your point and often have it listened to!

Burns[7] points out that there are only three possibilities when stating your point of view: (1) You are completely correct; (2) You are partially correct and the critic is also partially correct; and (3) You are wrong. He suggests the following methods to state your point of view:

You Are Completely Correct

Here, you diplomatically state your point of view. If the person continues to disagree, keep repeating yourself until he or she tires out.

> Critic. "I don't like the fact that you store your microscopes on this table."

> You: "I store the microscopes here so they will be accessible to the technicians. They don't have keys to the back room."

> Critic. "Yeah, but they should be stored somewhere else."

> You: "I understand your concern. I store the microscopes here so they will be accessible to the technicians."

> Critic: "Microscopes should not be kept out front."

> You: "You certainly show a lot of concern for the equipment, and that's great. I store the microscopes here so they will be accessible to the technicians."

By going over and over on your point, it is likely that your critic will rapidly tire. I'm tired from just writing the above exchange!

Burns suggests that when possible, state your point of view with an acknowledgment that you might be wrong or in some other way that allows your critic to save face. For example, you might be certain the environmental laws state that the person you are talking with has to have a permit to build a dock on her shoreline. Using a face-saving comment you might say, "I seem to recall that the county ordinance states that you will need a permit for this dock, but I might be wrong. Let's check the regulations and see for sure." Then you check the regulations and find that indeed the person needs the permit for the dock. This type of statement allows the person to save face. Allowing a critic to save face when you can is extremely important to their self-esteem and often allows them to comply more easily with any request you might have.

You Are Partially Correct and the Critic Is Also Partially Correct

Here you can negotiate an outcome using the negotiation techniques presented in this book (chapter 6). However, you have already presented yourself favorably to the critic by using the understanding and the empathy/agreement/strokes portion of verbal judo, which will help you constructively negotiate. For my current position as a professor I tried to get an electrofishing boat to start my lake studies in Arizona. I was negotiating with my future boss, who stated that the program just did not have the money to provide something like that, and hoped I would take the position anyway. I had already gotten favorable replies on other equipment and a higher salary. Instead of continuing to fight for the boat, I stated, "I certainly can appreciate that budgets are tight right now, and I really appreciate that you have worked so hard for a higher starting salary and other equipment for me. I certainly know how tough it can be working though all the government regulations on these things. Thanks very much! (empathy and strokes). I took the position, moved to Tucson, and about three weeks later, my new boss called me up to tell me they raised enough money to purchase an electrofishing boat for my program.

You Are Wrong

Assertively and quickly thank your critic for pointing out your mistake and apologize for any hurt you may have caused. Everyone is human and makes errors, and if you rapidly admit yours when they occur, you can impress your critic and others. During a workshop on freshwater fisheries management, a distinguished professor taught us a method to sample lakes. I spoke up to say I disagreed with the method and voiced my opinion that it was not statistically correct. The professor was very diplomatic but stated that the method was in fact valid. After class, I checked with some of our agency statisticians and found out that indeed the method was valid. I immediately called the professor and said, "Dave, I'm going to have to eat some crow big time! I was wrong about your method!" This admission of error, I believe, impressed the professor and to this day we have an excellent relationship. We are currently involved in a book project together. In addition, he has recommended me to others when other types of projects have come up.

Yes, But. . .

Here's a tip. When using the agreement strategy, try to minimize using the combination of words "yes, but." For example, "*Yes* you're right, you have modified your report quite nicely, *but* I think there are several parts that could be rewritten." Instead, when it is not too awkward, try substituting the word "and" or do not say anything at all during the transition. For example "*Yes* you're right, you have modified your report quite nicely. I think there are several parts that could be rewritten." Often if you use "but" after agreeing, you just negate the agreement.

Verbal Judo in Action—The Angry Man in the Truck

Now I will relate how I dealt with the angry man from the truck and the rest of the crowd at the lake where we wanted to sample. Following this actual exchange the crowd calmed down and we could continue our sampling. We were able to sample the lake for two years following this incident and made friends with many of the lakeside property owners. One was even a chief petty officer on a submarine who gave my crew and me a tour when the ship was in port! Another worked on antique steam locomotives for a hobby and we got to know him quite well. Immediately following the exchange, one of my research unit personnel was so impressed by how quickly the crowd calmed down; he told me, "Man, I'll follow you anywhere." Two instigators of the incident calmed down and went home, but they were not supported by the majority of the crowd any longer. The instigators said, "You slippery fish—you certainly talked your way out of this one!" While all incidents do not turn out like this, I was very proud of what happened. I did not have to back down, we were able to sample the lake, and the situation did not escalate. Using similar tactics may give you a way to work out of these types of situations. Here is the exchange that ensued.

I offered my hand to the angry man as soon as he got out of his pickup truck. He refused to take it and continued to yell at me. He was calling me and my agency names and was in general extremely angry.

Scott: "I'm here to listen to anything you have to say, sir" (understanding and allowing the person to vent).

Angry man: "What the hell are you doing here? This is a private lake and you have no right to be here. We don't want you here screwing up our lake" (he went on and on for about a minute or two, and I just let him go).

Other members of the crowd: "Yeah, just what the hell are you doing? What right does Fish and Wildlife have to push us around and do whatever they want to our lake? We don't want you here!"

Scott: "Folks, I can certainly appreciate that you are concerned with what happens to your lake. I definitely know where you are coming from. My mom and dad own a lake in Indiana, and you can bet they have a big say about what goes on there" (empathy). "Why don't you want us to sample fish in your lake?" (inquiry).

Another member of the crowd continued to go on: "Why are you showing up after we told you we didn't want you here? We told you we wanted you to stay away!"

Scott: "I can understand that you're pretty angry that we just showed up" (empathy). "I tried to contact you all time and time again to present to you what we would be doing at your lake. However, none of you would let us come out to talk with you. We finally worked with the folks on this side of the lake, and they are going to allow us to launch the boat over here."

Angry man: "What good would it have been for you to show up and give us a talk? You would have just had your government 'yes men' planted in the audience to agree with you and would have come out anyway whether we wanted you here or not."

Scott: "Okay, I know you're concerned" (empathy), "but here I'm standing, all alone, and my coworkers are in the truck. I don't have any government yes men here. You can ask me any question you want. Can you tell me specifically what you folks are so concerned about?" (second inquiry).

Other members of the crowd: "We heard about your study. You're just going to sample fish on this lake so you can poison it to kill all the fish that are eating the salmon."

(We have just gotten to the root of the problem—they think our team is here to poison their lake to kill the fish that are eating the salmon. They heard about the Lake Davis incident [see chapter 2] and they wanted to prevent the water in their lake from being treated.)

Scott: "Folks, you are all to be commended for caring what happens to your lake" (strokes). "We're here to investigate what the fish are doing in your lake. We picked your lake, not because we would like to poison it, but because there is a fish trap five miles downstream that can record the number of salmon leaving. Traps are extremely uncommon on lakes in western Washington. We want to find out what is going on with salmon—introduced fish interaction with western Washington in general, not your lake in particular. We aren't government agents in black helicopters. We're more like Marlin Perkins (*Mutual of Omaha's Wild Kingdom*, an old TV show that was kind of like *The Crocodile Hunter*). We're just like you" (empathy). "We want to protect the environment for your kids and grandkids, but to do so we really need your cooperation. Can you help us out? We have the legal right to work on your lake, but I really want to work with you folks, not against you" (diplomatically stating point of view).

People in the crowd: "Can you promise us no one from the State Fish and Wildlife will poison the fish out in our lake?"

Scott: "Again, I certainly understand your concern about having piscicide or some other chemical dumped into your lake" (empathy). "I cannot promise that your lake will not ever be treated for introduced fish. I hope you'll understand it's a large agency, and I don't make policy for it. I suggest that if that time ever comes, and we're certainly not here to do it, you take it up with the agency at that time" (diplomatically stating my point of view).

One person in the crowd to angry man (Bud) and female instigator (Val): "Val and Bud, I can't see any problem with what they are doing. Let's let them go ahead."

Another person in the crowd: "Okay, I guess it was just a misunderstanding. But we don't want Fish and Wildlife treating this lake with chemicals."

Scott: "Okay, I completely understand" (empathy). "We certainly don't have any plans to do that."

Now the crowd starts to disperse. A couple of interested homeowners come over to the electrofishing boat and we show them how it works. Val and Bud then come up to me.

Val: "You talked your way out of this one, you slippery fish. We'll be keeping our eye on you."

Scott: "I wish I could change your mind that we're really not here to do anything to hurt your lake. However, if you have any questions, please give me a call. We really do want to work with you" (diplomatically stating my point of view).

As you can see, there was not a rigid order of (1) understanding; (2) empathy/agreement/strokes; and then (3) diplomatically stating your point of view. While this order often works best, I used whatever communication technique or combination I thought was appropriate when each question or insult was aired. This flexibility allowed me to choose the appropriate strategy depending on where the crowd was coming from.

COMMUNICATION IN CRISIS SITUATIONS

What if the crowd had not dispersed as easily as it did? Law enforcement officers have developed communication techniques that work under crisis situations. Here we will discuss a few tips from their profession.

Gundersen and Hopper Techniques

In their book *Communication and Law Enforcement*, D. F. Gundersen and Robert Hopper suggest three ways to avoid crises when tempers are escalating out of control.[8]

Cooling Off

This is the extreme action of separating the warring parties. In domestic violence incidents police officers immediately separate the husband and wife to prevent further violence. In barroom brawls, we are all familiar with people separating the two combatants. Separating the parties allows them to "count to ten" or cool off for a while before dealing with the problem. If the crowd in the lake incident would have gotten angrier, despite my best efforts, we could have gotten back in our trucks and left the area. This would have been less preferable to working the problem out at the lake, but if the temper of the crowd was too high, it would have allowed everyone time to calm down before we tried another way to work with the landowners.

Smoothing

This technique uses compliments, stroking, small talk, and the exchange of pleasantries to calm the parties. Often the combatants are told what they have in common with each other. We conducted some smoothing tactics at the lake. We took every opportunity to show the landowners how much we were like them. We told them they cared about the environment, just like us; we all wanted to ensure our kids would grow up in a healthy environment; and I said my parents lived on a lake just like them. We talked and joked with them even though the underlying tension remained present throughout the night of that first sampling trip. We showed some of the landowners the electrofishing boat and described how we caught fish. We always tried to talk with them with an upbeat attitude.

Captain Mbaye Diagne—
A Master of Smoothing

In Rwanda, 1994, eight hundred thousand Tutsis and moderate Hutus were slaughtered by Hutu extremists in a genocide that lasted one hundred days. This was the fastest rate of mass killings in the twentieth century. Roughly 10 percent of the entire population of Rwanda was murdered. The United Nations was ordered not to intervene and the United States and other Western nations refused to get involved. Against direct orders, Captain Mbaye Diagne, a Senegalese army officer and unarmed UN military observer, took action. He saved the two sons of the prime minister from slaughter by hiding them in a closet as their mother was being killed, and he and other UN observers brought hundreds of Tutsis, moderate Hutus, and others to safety. Getting these people to safety was an almost superhuman task. In one instance he found twenty-five Tutsis hiding in an extremely dangerous neighborhood in Kigali. He successfully ferried them to UN headquarters in his jeep, five at a time past twenty-three checkpoints, each manned by Hutu militia killers. He convinced the killers at each checkpoint to let all the Tutsis live.

How did he do it? Mbaye was later killed in a random mortar attack and was not interviewed about his technique. However, witnesses say his ability to charm his way past the killers was uncanny. He would smile, joke with them, give them cigarettes, and make them feel confident. Belgians were targeted for killing in Kigali because they were considered by the extremist Hutus to be pro-rebel. A BBC journalist and Mbaye were stopped in their car at a Hutu extremist checkpoint. One of the militia members leaned into the car, waved a Chinese stick grenade under their noses, and asked if the BBC man was a Belgian. A wrong answer

(continued)

Captain Mbaye Diagne
(continued)

would have probably resulted in death for everyone. Mbaye joked with the militia member and said, "No, no—I'm the Belgian. I'm the Belgian here, look—black Belgian." That joke broke the tension of the moment, and the militiaman relaxed his guard. Then Mbaye said in fact the guy was BBC and had nothing to do with the Belgians. Because of the break in tension, the militia allowed the vehicle to pass.[9]

By smoothing his opponents with small talk, smiles, and jokes, this tall officer with the broad toothy grin saved hundreds of lives, even though he was unarmed and had no real power. These techniques have continually helped people talk their way out of dangerous, volatile situations. Smoothing techniques are also extremely beneficial for the conservation professional when working with angry ranchers, farmers, developers, landowners, and members of other agencies. Sometimes it only takes a joke, smile, or small talk to break the tension and allow for constructive problem solving.

Reframing

Often you can put a problem into a new light, so solutions that were not readily apparent before can now be identified. When two people are quarrelling with each other over a park management strategy, they can be told that the intensity of the fighting shows how much each of them cares about the environment. The situation is "reframed" and the parties look at it in a new light. In the incident at the lake, there was one lady who was flanked by her two children about ten to twelve years old. The lady was carrying signs against big brother (i.e., the government) working on her lake, and was showing her anger to us along with many of the other people. I reframed the situation by indirectly asking her about the example she was setting for her children. My statement "How can we protect the environment for your kids without your cooperation?" was directed at her. It allowed her to look at the confrontation in a new light, as a mother not setting a very good example in front of her children by accosting people who were concerned for their quality of living and the environment. After this statement she put down her sign, and they quietly went home.

Thompson Verbal Judo Methods

George Thompson was a former professor of English literature and a police officer. He analyzed interactions between adversaries, many of them between police and suspects, and advanced techniques where words were used to disarm adversaries to reduce use of guns, nightsticks, or mace. He has spoken to thousands of police officers across the United States about these techniques and was one week away from training four of the officers who were videotaped in the Rodney King beating. He claims that if these officers had taken his course, the beating may not have happened. Thompson recommends a five-step hard-style approach for handling a tough situation.[10]

Ask
If they are threatening you, or doing something you want to change, ask them to change their behavior.

> You: "Please do not take that prehistoric pot from that ruin."

Set Context
Explain to them why you are asking them to do this. Often our parents made the comment "You need to do this because I told you so!" However, both parents and agency personnel can often make people mad by taking this position. People have a deep desire to know why they are being asked to do something. In his police work, Thompson found that about 70 percent of people would comply after they were told why they were asked to do something. For the above example, a typical response might be:

> You: "If you take the pot from the ruin, other people won't get a chance to enjoy it, archeologists won't be able to study it, and you will be violating the law!"

Present Options
If they still seem like they do not want to comply, you can present them with options and the consequences of those options. The task of choosing an option is left with them. Allowing them to choose an option can often encourage compliance much more effectively than giving them a single order.

> You: "You seem like you still want to take that pot. You have a couple of options here. You can continue to take the pot and violate the law. It's a nice pot and probably *would* look good on your shelf. However, I and others here would have to report you and you could face a fine up to five thousand dollars, and potentially have to serve some jail time. You would likely lose your job for a felony conviction, lose a lot of money, and damage your relationships with your family, all for an old pot. The other option is that you put it back and go about your business. You don't spend any jail time, continue to have a nice wilderness experience, and save money and embarrassment. Which way would you like to go with this?"

Thompson is also a great believer in empathy. Statements like "Hey, I know exactly how you feel, that pot would look great on your shelf, wouldn't it?" (empathy). "However, is it worth all the trouble?"

Confirm
If the person still is insistent on his or her action at this time, you can give him or her one last chance to comply. Thompson recommends the statement "Is there anything I can say or do at this time to earn your cooperation?"

Act
Before you get to the act stage, most people would have done what you would have asked. However, if the action continues, rapid, decisive action is needed. Here without any further statements, you report the person for stealing the pot. Figuring out how you are going to act is important before you reach this stage.

DEALING WITH HECKLERS—
A COMMON OCCURRENCE FOR
CONSERVATION PROFESSIONALS

When I was a graduate student, I had to give a talk in an auditorium in front of a large national audience about the subject of my dissertation, grass carp. Grass carp are large plant-eating fish that are used for plant control. I was happy with my talk and when I reached the end, I asked for questions; a scientist with a national reputation for expertise in grass carp blurted out, "You're wrong. Grass carp don't behave like that!" I was crushed. Even though I had carefully designed my experiment and collected my data, I felt that I was a failure, and now so did everyone in the auditorium. It was one of my earliest dealings with a heckler in scientific meeting.

Since that first incident, I have dealt with numerous hecklers. They seem to be more of the rule than the exception at large public meetings. So having the ability to deal with hecklers is an important skill for any conservation professional or government official.

We are all familiar with how comedians handle hecklers. I remember Steve Martin's *A Wild and Crazy Guy* concert on tape. It was a huge hit when I was a teenager in the 1970s. In it Steve Martin is telling jokes to a large audience and you can hear a heckler in the back screaming something unintelligible. Martin ignores the man at first and then he quips "Yeah, I can remember my first beer, too." It shuts the heckler down and the audience gets a huge laugh.

In public meetings, the heckler can often be someone you will have to work with later, or someone in one of the groups that you are working with to try and get support. Therefore, humiliating the person in front of the audience, as tempting as it may be, will probably not serve your ultimate interests. According to David Burns, most hecklers share the following attributes.[11] They are intensely critical, but their comments are often irrelevant or inaccurate. They are often not well accepted among their peers, and their comments are expressed in an abusive style. Burns gives a short technique for handling hecklers that has often worked quite well for me. First, thank the heckler for the comments. Acknowledge that the points he or she brought up are important and emphasize that there is a need for more knowledge. You can encourage the critic to pursue meaningful research of the topic, or in the case of someone who goes on and on, you can encourage the critic to further share points of view after the meeting.

If you are often heckled, you might want to examine your speaking style to see if you might be contributing to the problem. Have you worked hard to make your presentation interesting? Are your points relevant and well thought out? Careful preparation and consideration of the audience's interest level in what you have to say can minimize heckling.

When the grass carp scientist told me I was wrong, I was inexperienced in handling hecklers. I asked him first to clarify what was wrong about my points. He said that grass carp in his study preferred to eat plant species in a different order than what I found in my study. I said something like "that's interesting, our data revealed something different." After my talk, we continued our conversation. Although our first couple of conversations were somewhat heated, it lead to further research investigating the effects of grass carp size and water chemistry of the lake on grass carp feeding preferences. We found that preference could indeed vary somewhat among sites and fish of different sizes. Later when I found out more about verbal judo, I gave the scientist compliments about his work. He opened up considerably after this, and I came to consider him a friend and an excellent source of information. This incident provided a valuable lesson in keeping my cool, acknowledging that other data might be correct, and giving compliments even in the face of criticism. Incidentally, I thought I would be humiliated in the eyes of my fellow scientists because of his comments. Interestingly enough, when I spoke with several of the other scientists following my presentation, they were actually angry at *him* for trying to embarrass a graduate student.

CONCLUSION

Used correctly, words can save lives, protect natural resources, make friends, and build reputations. I am unfamiliar with studies that have tested the effectiveness of these techniques in a statistically controlled manner. However, I have tried many techniques in communication and have found those discussed in this chapter to be among the most simple and powerful. Using these techniques for over twenty years has provided me with powerful, positive results, and I have also seen them successfully used by many other conservation professionals. Add these tools to your toolbox of communication techniques, and see what results they provide you.

CHAPTER SUMMARY

- Verbal judo can be a powerful technique to turn a critic into a supporter. It consists of (1) *understanding* your critic's point of view by asking questions; (2) using *agreement* in some form, *empathy* or *strokes* to diffuse the situation; and (3) then diplomatically stating your point of view.
- In volatile confrontational situations, cooling off, which consists of separating the parties; smoothing, which consists of using small talk, jokes, or an exchange of pleasantries; or reframing, which puts the problem into a new light, have all been used successfully.
- Another method of deescalating high-conflict situations consists of (1) asking them to change their behavior, (2) explaining why you are asking them to change their behavior, (3) presenting them with options they can choose and the consequences of following your requests or not, (4) asking them one more time to follow your request, and (5) taking whatever action is needed.
- One can deal with hecklers by (1) thanking them for their comments, (2) acknowledging that their points are important, and (3) encouraging them to conduct meaningful research on the topic, or asking them to discuss it with you further after the meeting.

CHAPTER FOUR

How to
Persuade People

I was standing in front of the Education Section of the American Fisheries Society National Meeting in Madison, Wisconsin. Sitting in front of me were some of the best fisheries educators in North America. A couple of hours earlier, in another room, I had just convinced the members of the Fisheries Management Section that a book standardizing fish sampling techniques across North America was a good idea, and they unanimously voted to support it. Now I wanted to sell the idea to the Education Section and get much needed support and money for the project. After my success with the Fisheries Management Section, I was relaxed and confident as I approached the front of the room where the Education Section had gathered. First, I described the purpose of the book: a reference authored by almost fifty sampling experts to recommend methods to sample different types of waterbodies in North America. At the end of my explanation, a hand in the audience went up. "Don't you think recommending a specific sampling method would have legal implications that you don't want to get into?" Another hand went up. "Do you want us to support this book, even though we haven't had much time to think about it?" And then another hand.

"Why do we need another book to do this? Can't we just add a couple of chapters onto an existing text?" I answered the questions to the best of my ability, and then asked if they would be interested in supporting the project. I was met with silence, and I felt like I was dying on stage. A vote was then taken to support the project, and the motion barely squeaked by.

What a difference from the Fisheries Management Section! Their support for the project was unanimous, and it had passed with one of the largest grants in the history of the section. But after I left the Education Section room, one of the other scientists came up to me and said, "Man, that was ugly!" I realized I did not do a very effective job persuading people in the Education Section.

The ability to persuade people to behave in environmentally friendly ways is at the heart of conservation. At the Washington Department of Fish and Wildlife, a respected older biologist commented to me, "Scott, we already know why the salmon are declining. It isn't a biological problem. The problem is getting the people to stop overfishing, stop destroying their habitat, not stock exotic fishes that prey upon them, and leave water in the rivers. Salmon declines are a social problem." He was right. The importance of the "people" factor has always been crucial to conservation professionals. Anglers and hunters are convinced to follow regulations, policy makers are persuaded to protect specific rivers or forests, and granting agencies are influenced to fund particular projects. Numerous texts and articles are available to conservation professionals demonstrating methods to investigate public wants, such as how to conduct creel surveys, determine angler and hunter wants and needs, and query various segments of the populations to understand their priorities. However, a discussion of techniques for persuading people is rare in the conservation literature.

Skills for persuading people or marketing a new idea are rarely part of training, but are necessary for successful careers as scientists, biologists, or managers. Some people discourage natural resources professionals from learning how to persuade people, declaring that they should not practice "advocacy." However, unconsciously or not, natural resources professionals attempt to persuade people almost daily—it is a necessary fact of life. It is not that conservation professionals try to persuade, but the *topics* on which they try to persuade, that make the difference between a behavior that is generally accepted and one that is not. For example, someone employed by the Sierra Club will have political advocacy as a key part of their position and they will attempt to persuade people to support various candidates or laws. Someone working as a research scientist in a national laboratory will often stay away from advocating various viewpoints or political candidates, concerned that such actions will damage her scientific credibility, yet she might try to strongly influence people to support a new direction of scientific research or to fund a new laboratory. Law enforcement personnel may not try to influence scientific direction, but will work to convince people to follow hunting regulations, or keep their cattle out of a creek.

You and other members of your specific occupation are best able to decide when advocating an idea is appropriate. Of course, persuading or influencing people in an underhanded manner, dishonestly using certain facts and hiding others to push your point of view at all costs, is never correct. However, using influence strategies to effectively educate people that using environmentally responsible practices can best meet their needs and wants is often proper. Additionally, if people on the "other side" of an issue are using influence strategies and you are not, you are at a distinct disadvantage when presenting the environmentally friendly position.

Methods for persuading people have been primarily developed in the fields of psychology, politics, and marketing. Strategies to persuade are used everyday to get people to buy soap, vote for a politician, or attend a baseball game. Let us investigate some of the basic ways people are influenced and how these might be used in conservation. Remember, you do not have to use these if they make you feel uncomfortable; however, they can often dramatically increase your effectiveness.

NEEDS OF PEOPLE

Abraham Maslow (1908–1970) was a psychologist who was one of the founders of humanistic psychology in the mid-twentieth century. Perhaps Maslow's most famous accomplishment was the development of a "hierarchy of needs."[1] This simple pyramid has been used to shape policy and influence marketing since its inception in the early 1940s (figure 1). Maslow's theory states that people have specific needs that are arranged in a hierarchy, with the more basic needs being met first, and after the basic needs are met, higher needs are tackled, moving up the triangle. For example, physiological needs must be met first, including breathing, eating, and sleeping. Once these most basic needs are met, safety becomes important, then social acceptance, and self-esteem. The highest state that can be obtained in the hierarchy is self-actualization, a feeling of "having arrived" or accomplishment of most of life's major goals. According to Maslow, only about 2 percent of the population achieves a level of self-actualization. Higher needs become important only after the needs lower on the triangle are met.

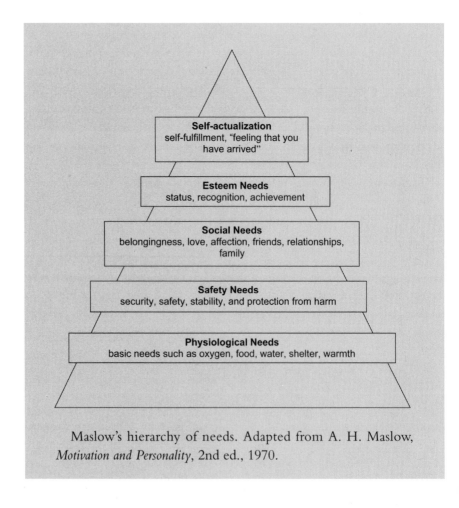

Maslow's hierarchy of needs. Adapted from A. H. Maslow, *Motivation and Personality*, 2nd ed., 1970.

By knowing where the audience's needs are on the hierarchy, the manager can adjust his or her arguments or solutions to meet those needs. For example, a manager trying to convince an audience from an economically depressed logging town that cutting restrictions and stream buffers are needed to protect the biodiversity of stream fishes will probably be met with hostility and anger. An underemployed logger's needs are at the physiological and safety portion of the triangle, while an argument to protect biodiversity appeals more to those who are in the self-actualization, esteem, and social levels of the triangle. A more effective argument to convince these individuals might be how excess logging will not make it possible to meet their physiological and safety needs, mainly providing food and shelter for themselves and their family. For example, cutting trees faster than they are replaced and not leaving stream buffers will result in your continuing to be out of work, because all the trees will be gone. It will also affect many other money-making industries in your town such as ecotourism and angling, which will probably hurt your ability to make money as well. Therefore, it is in your best interest, in the long run, to leave buffer strips and lower your rate of cutting to a level the forest can support. This will ensure you will keep your jobs.

In the late 1960s and early 1970s alligators were vanishing at an alarming rate from the swamplands of the southern United States. One of the factors contributing to their decline was huge illegal poaching operations.[2] Federal and state agents believed that a wide variety of poachers, crooked fish and game agents, retailers, and exporters were involved. Law enforcement went to work—new federal laws and aggressive enforcement reduced the number of buyers for the hides, devaluing them considerably. To infiltrate the poaching operations, U.S. Fish and Wildlife Service agents collaborated with those who were connected with the business. Some of the low-level poachers, many from Cajun backgrounds, initially started to cooperate with law enforcement out of fear of their higher-level partners. If you knew too much, you could disappear in the swamps. However, even though they initially came in only trying to save themselves from being killed, after talking with dedicated and persuasive U.S. Fish and Wildlife Service interrogators, these Cajun hunters often turned into ardent environmentalists. How did this happen? According to Mark Reisner, generations ago these Cajuns were pushed onto lands that were not much good for anything but hunting and fishing. When game started to disappear, many of the Cajun hunters just kept on hunting, finding ways around the law. Instead of arguing that the alligator would go extinct, the Fish and Wildlife Service agents contended that the way of life of the Cajun bayou residents would come to an end if the hunting continued unchecked. Hunting and fishing were great, they argued to the Cajun hunters. The United States of America was a free country, and the agents wished they could go back to the times when the fish and wildlife were abundant and everyone did as they liked. However, they argued, things weren't like that now. If the Cajun hunters and others kept poaching at current rates, there would no longer be wildlife, such as alligators or ducks, to provide food for them or to support their way of life. The agents successfully converted some of the poachers into conservationists, not by discussing the merits of biodiversity, but by realizing the poachers were on the lower end of the triangle, and convincing them that their very existence and life in the swamps would be altered if they did not manage their game. These converted poachers were often sent out on undercover operations designed to protect the alligator. The successful protection of the alligator from poaching and overhunting, through strong regulations and effective enforcement, is regarded by many as one of the factors contributing to their high populations today.

For somebody much higher on the triangle, say a wealthy owner of a large logging company, appealing to his or her ego or social needs might achieve better results. For example, convincing them that overcutting would result in a considerable amount of bad publicity for them and their firm might be an effective way to influence them. Conversely, if they set aside areas along creeks and rivers for riparian areas, it would benefit them with good public relations. Perhaps other creative ways could be used to appeal to their needs, such as naming a park after them, or presenting them with an award at the town council or public meeting. Determining where the person is on the triangle and using an argument that will help them meet the appropriate need is often quite effective.

TECHNIQUES OF INFLUENCE

Advertisers have a long history of influencing people using methods based on human psychology. Arizona State University social psychologist Robert Cialdini spent years studying how people are influenced. Early in his career, he attended numerous seminars designed to teach sales staff how to persuade people to buy things. From these seminars and his subsequent research, he learned that most people react automatically to events. This allows them to

Don't Think People Can Be Influenced About the Importance of Endangered Species? What About These Popular Items?

Some biologists have told me there is no way we can get a majority of people interested in saving an endangered plant, salamander, insect, or snail. While most conservationists feel that rare species provide priceless information and diversity to future generations, some argue that most people just will never care about such "insignificant" organisms. Perhaps it is just poor marketing of the value of endangered species. Look at the "must-have" things that millions of people have clamored, sometimes rioted, to buy. Usefulness is most definitely in the eye of the beholder and is often the product of a good marketing campaign.

- Mood rings
- Pet rocks
- Cabbage Patch dolls
- Pokeman
- Tickets to a Beatles concert
- Humvees
- A cheese sandwich that "bore the image of the Virgin Mary" (Sold on eBay for $28,000 in November 2004)[3]

successfully deal with the large amount of information they receive each day. An event happens, it is processed in a similar manner by a remarkably large percentage of the population, and there is an automatic response. Advertisers have learned that they can achieve an automatic response from people, simply by knowing a few simple principles of how people are influenced. Most successful natural resources managers, knowingly or not, often use these influence techniques. Cialdini discusses six methods to influence people.[4]

Liking and Similarity

This theory states that we are more willing to be influenced by those we know and like. If we like someone, that person is able to convince us and influence us to do things we may not normally do. For example, a friend is often able to convince us to try something new, such as ask someone out on a date, or invest our money in a specific way, while a person we did not know or like would be unable to convince us. I remember seeing a television special on a reformed racist and white supremacist. He was serving prison time on an unrelated offense, and got involved with a white supremacy group while in prison. He had shown no previous inclination toward white supremacy. He stated that the reason he joined was Christmas cards! Evidently the white supremacist group was the only one who treated him nicely and was interested in him, even to the point of sending him Christmas cards. This influenced him to join their group and support their cause. In natural resources management, the manager who gets along well with the angling club or the homeowners group is usually much more successful in getting them to do something than the faceless bureaucrat whom they neither know, nor like.

Similarity is grouped with liking. We usually like people who are most similar to us. For example, most of us know the stories of how Abraham Lincoln, with his rural twang and mannerisms, could convince a country jury that his client was innocent more easily than a slick lawyer with a clipped urban accent and dapper dress. In the natural resources profession, similarity is used often to influence constituencies. I know a southwestern fisheries biologist who never travels into rural ranchland areas without wearing a cowboy hat, Wrangler jeans, and carrying a container of tobacco in his back pocket. Any rancher he meets thinks he is one of "them" even though he is a talented environmental manager supporting environmental issues with which many of them might disagree. This same individual goes to urban meetings and conferences dressed up and polished. How similar we are to those we need to work with often can influence our effectiveness.

Conservation professionals can use the liking and similarity principles by working hard to develop positive close relationships with our constituencies, politicians, other agency members, and funding agencies. Most people do things well, so identifying them and complimenting the people can increase your influence with them. Compliments are easy, fun, and make both you and the other person feel good. I find that if I have to face a hostile audience, beginning the talk with a compliment (e.g., "It's great to see so many of you here—you should be commended—most people wouldn't show enough dedication to give up their Friday evening to come out and talk with me") makes it is much easier for the evening to go positively.

Another powerful way to develop liking is for people is to work together toward a common goal. For example, suppose you are an environmental regulator. You will often be able to influence landowners more effectively by working with them whenever possible "on the same side of the table," with the common goal of enhancing their property values or making their land more livable for them and their family, rather than working "across the table" from them, handing down fines and forcing them to do certain things. In Pima County, Arizona, developers, environmentalists, academics, and government officials worked together on a conservation plan for the Sonoran Desert that would allow open space in the face of rapid population growth. By collaborating, these people developed the ability to influence one another in a way that would have been impossible as antagonists.

Authority

The authority rule states that people will be influenced more by persons whom they perceive are authorities on subjects. Cialdini discusses an experiment on human reaction to authority that was conducted at Yale University from 1961 to 1962.[7] In the Milgram experiment, college students were recruited to give a subject electric shocks on orders from an authority figure in a white laboratory coat. Even though it distressed the college students greatly, most shocked the subject at the highest levels when the authority figure asked them to do it, even though the subject was apparently in much pain. Conversely, when the subject asked for higher and higher electric shocks, and the authority asked the student to stop, the student did stop. In all studies, the subject was an actor and was actually in no pain. The experiment vividly demonstrated the power of suggestion by an authority figure.

Authority figures can be a powerful means to convince people to take action in the natural resources area. Celebrities are often used as spokespeople

John Muir—Mr. Personality

When one thinks of legendary conservationist John Muir, what often comes to mind is a gaunt, unsmiling man with a giant beard, silhouetted against the sky on top of a mountain peak in the Sierra Nevada. The last adjective that one thinks of to describe him is charming. However, Muir would not have been nearly as effective if his personality was not so engaging. To convince people of the importance of wilderness areas, Muir not only showed them the utter majesty of the Sierra Nevada. He turned out to be a highly charming traveling companion, and very convincing as well. Theodore Roosevelt, one of our most influential presidents regarding conservation issues, complimented him highly. He spent three days with John Muir in Yosemite in 1903, and according to his accounts, enjoyed every minute. Roosevelt wrote about Muir's personality: "Ordinarily, the man who loves the woods and mountains, the trees, the flowers, and the wild things, has in him some indefinable quality of charm, which appeals even to those sons of civilization who care for little outside of paved streets and brick walls. John Muir was a fine illustration of this rule."[5] Furthermore, he described Muir as one "brimming over with friendliness and kindliness," also stating "there was a delightful innocence and good will about the man, and an utter inability to imagine that anyone could either take or give offense." No doubt that Roosevelt was already a conservationist and did not need much convincing about the beauty of the Sierra Nevada. However, Muir's likable personality and similarity in thought on conservation issues helped reinforce many of Roosevelt's strong conservationist ideas. One day after leaving Muir and the Yosemite Valley, Roosevelt asked Interior Secretary Hitchcock to extend the Sierra Reserve northward, all the way to Mt. Shasta.[6] The liking and similarity principle helped protect a vast, priceless piece of America's wilderness.

for environmental causes, and even though few have received any type of biological training, these people are often effective because of their general perceived experience as "authorities." In the 1960s and 1970s controversy over who could catch salmon in Washington State ran high. Treaties from the 1850s stated that Native American tribes who signed could fish "in-common with" people of European extraction. This had huge implications for both tribal and non-tribal fishermen. Celebrities such as Marlon Brando fished with the tribal members to show their support.[8] Later tribal members recalled that the participation of Brando had been crucial to rivet national attention to the fishing controversy and garner support for the tribes.

Authority figures do not have to be celebrities, but often can be prominent figures in the hunting, fishing, scientific, or conservation communities. Many agencies and universities currently use advisory boards. These advisory boards serve many functions, such as providing outside advice to the agency, serving as a sounding board for new ideas, and helping to support programs within the agency. Because members of these advisory boards are often prominent members of their communities and perceived as authorities, they can often have significant clout when convincing members of their organizations about the usefulness of a management action.

Once, when I spoke in support of a state biologist at a public meeting, I received an ovation for my comments, even though I stated points that had already been discussed by state biologists. The biologist later asked me why I thought my comments were received so positively, even though I echoed a point made repeatedly by him before. Because I was a university professor in fisheries, I was probably perceived as an "authority" on the subject, even though I had far less experience and knowledge on the topic than the state biologist, and I was new to the area. Conservation professionals who work with people in authority in a community and get them to support their position can often influence large groups of people. Authorities can include politicians, civic leaders, or businesspeople; respected members of angling or hunting groups; leaders of conservation organizations; or prominent scientists. It is important that the authority be perceived as an authority by the group you are hoping to influence. For example, a leader of the Audubon Society or Earth First! would not be a good authority figure to convince an audience of loggers that harvesting old-growth timber was something that should be avoided.

The Power of One—
Finding the Authority Figure

Lyndon Johnson was our thirty-sixth president, but many claim his true genius was as a legislator in both the U.S. House and Senate. Johnson was remarkably effective, and was called by Pulitzer Prize–winning biographer Robert Caro "not only the youngest, but the greatest Senate leader in America's history."[9] Johnson became majority leader after a single term in the U.S. Senate, something unprecedented in this body that rewarded seniority. Johnson stated that "the way to get ahead is to get close to the one man at the top." One of the techniques Johnson used in his rise to power was to first identify the most highly influential members of the House and Senate, such as Speaker of the House Sam Rayburn, and the powerful leader of the Southern Caucus, Richard Russell of Georgia. He then made it a point to court these highly influential authority figures, both professionally and socially. They in turn helped Johnson in his rise to power. Today, if one works with legislators, committees, or groups, it is equally important to identify the most influential member of that committee and prioritize a good working relationship with them.

Reciprocation

One of the strongest human urges is to pay back or reciprocate when something is given to us. Advertisers will use this by giving out "free" samples, often making people feel obligated to buy. Hare Krishnas give free flowers to passersby at public facilities and ask for donations. After people receive the flowers, they feel compelled to give the Hare Krishnas a donation, even though they might be considerably opposed to the goals of the group.[10]

How can the reciprocation principle be used in conservation activities? Biologists who volunteer on other's projects are often able to convince the other biologist to volunteer on their project in return. When I worked as a biologist with the Washington Department of Fish and Wildlife, we always gave the first chance at job openings to those who had volunteered with us previously. Not only did we have a chance to get to know the volunteers and judge how well they worked, we also felt somewhat "obligated" to give them the first shot at the position, since they had first given us their volunteer time. For this reason, I strongly suggest students of mine volunteer whenever possible with biologists or conservation agencies for whom they might like to work.

Another way that the reciprocation principle can be used is to send people information or give them something to obtain their support. The Arizona Game and Fish Department has popular, nicely designed posters on the native fishes of Arizona. These posters, when given to the public, not only inform them, but because of the reciprocation principle, influence them to do good things for the department or native fishes.

Many times a deadlocked negotiation can proceed when one side gives something up. It does not have to be the crucial core points, but it can be something small. Because of the reciprocation principle, the other side is often compelled to give something up as well.

Commitment and Consistency

Another strong human urge is to be consistent in what we do and say. If we give a promise, we wish to continue to stick by it even if it is uncomfortable to us. One of the striking examples of the consistency principle that Cialdini gives in his book was the brainwashing phenomenon of Korean War prisoners of war. In the Civil War and in World War II, POWs were fiercely loyal to their cause, even when incarcerated in the enemies' prisoner of war camps. However, American POWs incarcerated in North Korean prison camps often came back as strong supporters of Communism. How did this happen? When first admitted to the camp the prisoner was not asked to divulge information. He was given frequent sessions on the benefits of communism, and later the soldier was asked to write about a few things he heard in the lectures that he liked about communism. The soldier figured he was not divulging any sensitive information, so he complied. After he wrote more and more things down on paper about what he liked about communism, he felt committed to his point of view. To remain consistent with his commitment, especially after it was written on paper, the POW felt the need to defend his previous words, even after he returned to the United States.[11]

There are many examples of how this principle works in the conservation arena. A fellow professor I work with always asked a granting officer for a verbal commitment to fund a project. After this, it was very uncomfortable for that person to later deny funding. Getting the person to verbally commit to an action was often sufficient motivation to ensure they followed through. The consistency principle is also what makes it so hard to change programs (or people!) that have been in place for a long time. "We have always done it that way" is the argument commonly made. In the past, angling regulations for a state would often fit on the back of a notecard. Now, an increasing human population, habitat modification, and a desire to provide a diversity of angling opportunities require biologists to manage waterbodies differently. However, a frequent refrain from fisheries management staff is that the "agency needs to shorten and simplify the angling regulations to fit on an index card like they did in the past," even though a list of regulations by waterbody may better manage the states' fisheries resources. Changes in largemouth bass length limits ran head on into the consistency principle. For years, minimum-size limits were used to manage largemouth bass populations. In this regulation, largemouth bass under a specific "minimum" length cannot be harvested by anglers. However, after many years of these regulations, large numbers of largemouth bass accumulated or "stockpiled" under the minimum size. The largemouth bass would compete with each other for food and grow slowly, so managers wanted to remove a few and allow the rest to grow. Therefore, regulations called "slot limits" were put into place to allow largemouth bass under the minimum size limit to be harvested, to protect medium-sized fish with lengths in the "slot," and to allow harvesting of fish larger than the slot. Many fisheries managers reported that anglers did not want to use the slot limit. It was inconsistent with the previous regulations. After years of not being able to harvest fish under the minimum size, it was very hard for anglers to then keep the fish.

How can you use the consistency principle to influence someone? The key is to get them to agree with you that they will do something about a subject, or that they practice an action so many times it becomes ingrained as part of their behavior. If it is a difficult action or one that they might disagree with in some way, get them to commit to smaller requests earlier, and work up to larger and larger requests. Cialdini illustrates this point by referencing an experiment from 1966 by psychologists Jonathan Freedman and Scott Fraser. In the experiment, homeowners were asked by researchers posing as a safe-driving organization if they would put up a small three-inch square card in their window that would support safe driving. Next a larger group of people, including those who had put the card in their window and others who were never approached in the first place, were asked to put a large, ugly, plywood billboard in their front yard reading "Drive Safely." Of those who put the small card in their window first, 76 percent agreed to put the large, ugly plywood sign in their yards as well. Only 17 percent of those who had not been approached earlier to put up the small card, and therefore had not agreed first to the small request, agreed to put up the ugly plywood sign.[12]

At all times, let people know how their action would make them appear consistent with their previous actions. For example, people who feed urban wildlife can often attract unwanted animals to their yard, resulting in negative human-animal interactions. A way of influencing these people is to tell them that their actions of feeding wildlife are bad for wildlife, which is inconsistent with their previous interest in caring for animals.

Social Proof

Why do bartenders often "seed" the tip jar with money? Why do movies often have the reviews printed on the box at the rental store? Why do fads exist? These are examples of taking advantage of "social proof." People are influenced to behave in a certain way if someone else is doing it. We all knew people in high school who seemed to get an enormous amount of dates. If a lot of people were dating this person, then the perception was that this person was more desirable, so, in turn, even more people wanted to date the individual. Psychologist David Burns called this the "Queen Bee Phenomenon."[13] The more people who do something, the more powerfully a person is influenced to do that very same thing. For example, if a lot of people feel that an angling regulation or an endangered species act ruling is correct, it is more likely that others will adopt this opinion as well.

Social proof is an extremely powerful force to influence people to do things, even to give up their own lives. The RMS *Titanic* hit an iceberg on a bitterly cold April night in the flat, calm north Atlantic in 1912. A large crumpling gash was smashed by the iceberg on the liner's side, and soon the ship started to go down by the bow. Lifeboats were readied, and the call was given for "women and children first." Social proof held most of the men back from entering lifeboats, even though some were launched half empty. Millionaire Benjamin Guggenheim helped women and children into lifeboats then dressed in his finest evening clothes to await his death. He remarked to Steward Henry Etches that "no woman shall be left aboard this ship because Ben Guggenheim was a coward."[14] However, J. Bruce Ismay, managing director of the White Star Line, helped people into lifeboats, and then when one was being launched less than full; he jumped in and saved himself. Even though no other women or children were around to enter the boat, and Ismay would have just been one more life lost if he had not gotten into the boat, he was vilified in the American press for violating the social rule of "women and children first." Called J. "Brute" Ismay and a coward by the Hearst press, he spent the rest of his life living with harsh criticism. Texas and Montana towns named "Ismay" seriously considered changing their names so they would not be associated with the disgraced White Star Line director.[15]

Social proof has been a powerful force in fish and wildlife conservation. Few issues in fish and wildlife make it to the U.S. Supreme Court; however, the case of the Devils Hole pupfish was one (discussed in chapter 1). A small, seemingly insignificant fish was protected not only because of court rulings, but also overwhelming public support. Scientists, managers, and the public worked hard to publicize the plight of the pupfish. Newspaper articles, bumper-sticker campaigns, and documentaries were produced about the fish. The story of the Devils Hole pupfish appeared in the 1970 documentary *Timetable for Disaster*. The producer-director of the documentary went to the secretary of the interior and said that if nothing was done to prevent the Devils Hole pupfish from going extinct, they were going to do another documentary on how federal agencies stood by and let the species become extinct. The threats of publicizing the loss of the fish helped increase action by government agencies fearful of bad publicity. The application of social proof, and legal work through the U.S. court system protected this species from extinction.[16]

How can you apply the social proof principle in conservation? Work hard to get as many people as possible to support your position. For example, if I submit a proposal, I try to include support letters from as many different agencies and people as possible. These letters of support can convince reviewers of the proposal that the project is a popular one. This is a powerful way to get support from the reviewers. If you are trying to get the public to do something, such as follow a hunting regulation, dispose of litter properly, or support a bond issue funding a new water treatment plant, find some way to let them know that your idea is very popular with a large number of people, and work hard to get a large number of people to support it.

Social proof can also be used to convince people *not* to degrade your ideas and projects. My grandfather was mayor of a small Indiana town and worked hard to install a new water purifying system. Many people in the town were convinced that the water would taste bad and were ready for the date the treatment plant was turned on so they could call the mayor's office and protest. When the big day arrived, the water treatment plant operator called my grandfather to tell him that the plant could not be put into operation that day because of mechanical problems. He swore the plant operator to secrecy, and both of them swore all those in their families to secrecy as well. On the appointed day, his office received numerous complaints about how bad the water tasted, even though the new plant was not turned on. My grandfather just said, "Sorry about the taste of the water—we are trying our best to do something about it." Two days later, the plant was started up, and a big article appeared in the newspaper proclaiming the plant was now operational, two days after the projected start date. All those who had complained, did not dare to now because of the social proof principle!

Scarcity

Ever notice that if something is hard to get, it seems that much more valuable to you? If something is rare, it seems much more important? This is the scarcity principle at work. We seem to value things more if they are scarce. Think of all the advertising slogans that use this principle to get you to buy things: Order by midnight tonight or you might lose this offer! There are only a limited number left! This plate is limited edition! This offer is not available in stores!

According to many geologists, diamonds are actually a fairly common stone. Before the nineteenth century, they were exceedingly difficult to get, but with the expansion of mining operations, the stones became fairly

Arrogance—The Silent Killer
of Influence and Goodwill

During my career, one of the biggest killers of public support and agency cooperation I have seen is arrogance. Not confidence in one's own abilities, which can be a good thing, but an arrogant attitude that puts down the ideas and thoughts of others. A university professor might think that an agency technician doesn't know the latest "cutting-edge" technique or isn't quite as intelligent as she is. When I was an agency biologist, a professor who was talking with me left in mid-sentence to talk with a "more important" fellow professor. An agency person might think that all university professors are ivory-tower eggheads. When I was a university professor, agency types would often talk about academics "not living in the real world." And of course, academics and agency personnel can both be arrogant with the public they serve.

Most people can spot arrogance a mile away, and it can profoundly affect the ability of the arrogant person to get things done. One of my bosses used to call this affliction "Ph.D.itis" and said it could be devastating to a career.

If you see a landowner, angler, agency biologist, or administrative assistant and feel like being arrogant and throwing your weight around, realize a person's outward appearance or occupation might not reveal the true person. The following are true descriptions of some people who you might feel "better" than, and might be tempted to treat in an arrogant manner. Their names are at the bottom of the page.

A. Wore ill-fitting clothes, grade-school dropout, hired farm hand, clerk at a convenience store, several failed attempts at government employment, plagued by depression and insecurity, attempted suicide.

(continued)

Arrogance (continued)

B. Consistently used profane language, quit school in the seventh grade, fired from several jobs, divorced with three kids, arrested for assault and battery, involved in a shooting incident, pumping gas at age forty.

C. Failed numerous courses at school, father thought he was worthless, failed three examinations to get admitted to military school, defeated five times for political office, fat, could not speak loudly, had trouble saying the letter "s."

D. Teacher told principal that it was useless to keep him in school any longer—he was "addled," started a railroad-car fire, deaf, often fell asleep on the job, fired a couple of times, homeless, shabbily dressed, subject of ridicule by his coworkers.

(A) Abraham Lincoln,[17] (B) Colonel Harlan Sanders, founder of Kentucky Fried Chicken,[18] (C) Winston Churchill,[19] (D) Thomas Edison[20]

widespread. Of course a common stone would lose its value, and destroy the public's faith in diamonds, because a "rare" item is worth more. This is well known by the De Beer's Group, which sells about 90 percent of the world's diamonds, and according to some, also fixes the prices of the world's diamonds by hoarding stockpiles and limiting production. The De Beer's Group was charged in a price-fixing scheme by the Department of Justice in 1994, but the company failed to go to court. That is why you will not see them selling in U.S. markets such as New York, but concentrated in foreign cities such as Antwerp.[21]

In wildlife and fisheries, a dramatic example of the scarcity principle is shown by how we value similar animals and plants that only differ by how rare they are. Think about a turkey vulture. A red-capped black bird, circling and flying, looking for carrion. I like the turkey vulture, but there are many who would just as soon shoot it as look at it. Now contrast that with the California condor, extremely rare, also red head, dark body, enjoys eating dead things, but is bigger than the turkey vulture. Countless dollars have been spent on keeping it from the brink of extinction. So why is one so greatly prized, resulting in expensive propagation efforts, protests if they are in any way threatened, and also the subject of children's books,[22] while the other is pretty well ignored? Because there are less than one hundred California condors in the wild, while there are thousands of turkey vultures.

Another example comes from the fish world. The European or German carp is a highly prized fish in England, where it is a rare catch. Contrast that with the carp in the United States. Here the fish is extremely numerous and its sheer abundance make it a pest. The carp is an attractive fish, with large cycloid scales and a brassy shimmering body. The roundtail chub is another member of the minnow family, with smaller cycloid scales, a thin tail for moving through the water quickly, and a silvery, brassy appearance. There is not much difference between the fish in appearance. However, in the United States, the carp is considered an invasive pest, while the roundtail chub is a prized species of interest.

The scarcity principle also comes into play when people might lose something. People hate losing something much more than gaining some-thing. Therefore, an argument that you might lose something is often more powerful than arguing you will gain something. For example, I was at a research staff meeting when I used to work for a state agency. The depart-ment had some money to spend, and our supervisor was trying to deter-mine what program would get it. Asked about what things the department should spend money on, the head of the computer section said that a new computer system would help the agency gain efficiency in computer oper-ations. The head of the fish aging unit argued that a new microscope system would greatly improve scale reading capabilities. I was head of the Inland Fisheries Research Unit, and we had an old decrepit electrofishing boat used for catching fish. I commented to the group that the electrical system on our electrofishing boat regularly shorted out and had once caught fire. Someone could lose their life or we could lose a lot of money in medical bills if we didn't replace it. Because of the rarity principle, our unit was selected as the recipient of the funds for replacing our electrofishing boat. When we were subjected to budget cuts, the staff who were most effective at maintaining a program were those who told the administration how much they would lose by cutting the program, not those who argued how much would be gained if the program was continued. I was once able to protect my job in the face of huge budget cuts by telling my superiors how many tens of thousands of dollars would be lost in projects that were already underway if my position was cut.

The rarity principle is not only a technique for influencing people. If it were not for the rarity principle, we would not try so hard to keep unique areas, places, plants, and animals on the globe for our enjoyment and that of future generations. The rarity principle means that historic buildings, paintings, rare animals and plants, and unique habitats stand a good chance of being protected if we do a good job at emphasizing how rare and special they are, and making clear what will be lost if certain actions are not undertaken.

CONCLUSION

As a natural resources professional, you will possess valuable knowledge about the environment that most people will not have. It is up to you to learn how to present this knowledge in the most effective manner to make a positive difference. Use the skills presented in this chapter and you will better convey information to people, convince them of its importance, and show them how using environmentally friendly practices can best meet their needs and wants.

CHAPTER SUMMARY

- Maslow's hierarchy of needs and Cialdini's six compliance strategies both describe how people are influenced.
- Maslow's hierarchy of needs states that people must meet their basic needs, such as food, shelter, and water, before more advanced social and esteem needs are met.
- To influence people using Maslow's hierarchy of needs, try to find out where your audience is on the hierarchy, and then emphasize how your conservation strategy will meet that particular need. Arguments that show how the conservation strategy will meet a social need will not be

effective on an audience whose needs are more basic. The converse is also true.

- According to Robert Cialdini, the majority of people's behavior is automatic. Advertisers, marketers, and others use this concept and six major strategies to influence people: liking and similarity, authority, reciprocation, commitment and consistency, scarcity, and social proof.
- The liking and similarity rule states people are more willing to be influenced by those they know and like.
- The authority rule states that people will be more influenced by persons who they perceive as authorities on subjects.
- The reciprocation rule states that people have a strong urge to pay back or reciprocate when something is given to them.
- The commitment and consistency rule states that people have a strong urge to be consistent in what they do and say.
- The scarcity rule states that people value things more when they are scarce.
- The social proof rule states that people are strongly influenced to like things that other people like.

CHAPTER FIVE

Customer Service and Getting Funded

One of the professors who advised me as a graduate student was probably the best fisherman I had ever seen. He had worked on charter boats off the coast of California in the 1960s, regularly caught salmon in Puget Sound and Alaska, and was a talented freshwater angler. He took a trip with his students to Neah Bay to catch groundfish, and he had no problem quickly catching all the fish he was allowed. I was never much of a hook-and-line fisherman, but one of the other students was quite talented; however, he never got quite as good as Gary, the professor. He always asked Gary his secrets for catching lots of fish, but according to him, Gary was always hesitant to tell him, because Gary thought there were too many really good anglers around, and the resource could not handle a lot more.

Let's face it, many conservation professionals (sometimes including me!) can be hesitant about telling people how it is possible to get grants and other types of funding. It seems as if there is only so much to go around, so we do not want too many other people who are "good" at it. However, the very fact that you are reading a book like this shows you probably have the aggressiveness to get your programs funded. One can argue that good science and conservation programs help the economy, which in turn provides more money for science, which helps the economy. According to that logic, sharing with you a few "tricks of the trade" will probably not take too much of the pie from others!

Why does this chapter group "getting funded" with "customer service"? Because the key to getting funded (more than once) is providing excellent service to your customers. Many conservation professionals believe that customer service and other related subjects are best left to business people. However, you are most definitely in a "business." If you are a researcher, that business is supplying accurate, rapid research information to natural resources decision makers. If you are a manager, your business is working to provide outdoor opportunities to the public, and preserving the resource for future generations. If you are an educator your business is providing the highest quality of training for your students.

SECRETS OF CUSTOMER SERVICE

Who are your customers in the conservation arena? Of course the resource itself is ultimately what natural resources managers are interested in managing and protecting. Therefore, some biologists believe their customers are the fish, animals, or plants with which they work. However, fish, wildlife, and plants don't vote, support budgets, or influence people. The key to protecting and managing natural resources is working well with people. Therefore most argue that people, not the resources, are the customers. People influence budgets, programs, and departments. Good service to people results in strong budgets and well-funded programs, which in turn allow the biologist to do good things for fish and wildlife. Strong fish and wildlife populations and programs in turn please the human customers. They in turn support the fish and wildlife programs further. It can be a snowball effect.

Most everyone can name important customers. Of course your boss is always a customer. He or she depends on you supporting his or her role and advocating his or her programs. Other customers can vary depending on the position. As a professor, my customers are the people who use information that my students and I obtain. For me this includes state, federal, and local biologists and managers, and other university staff who depend on me to serve as a liaison to these agencies. My customers are also my students who depend on me to provide them a quality education. As a laboratory technician, your customers are the biologists, researchers, and managers who use your laboratory results. As a department head in a university, your customers are alumni, the faculty, and the community at large. As a politician, your customers are the people who vote for you (your constituents).

Providing poor customer service can cripple your ability to get things done. In Michael LeBoeuf's classic book on customer service, *How to Win Customers and Keep Them for Life*,[1] LeBoeuf contends that your greatest threat is the nice customer! Here's why: If you provide poor service, a mean, vocal customer will tell you what is wrong (usually quite forcefully!) and give you the opportunity to correct it. However, most dissatisfied customers will suffer in silence, never telling you, but they will not use your business again. A typical business only hears from 4 percent of its dissatisfied customers. The other "nice" customers (96 percent) just quietly go away, not wanting to antagonize anyone; 91 percent never come back.

If you provide poor service, the damage to your reputation is not just restricted to the dissatisfied customers. According to LeBoeuf, a typical dissatisfied customer will tell eight to ten people about their experience. One in five will tell twenty. Once the damage is done, it takes, on average, twelve positive-service incidents to make up for one negative incident. Therefore, it is essential to keep as many of your customers as happy as possible.

In the natural resources field, it is important to state what "keeping the customers happy" means. It does not mean allowing people to harvest all the fish or wildlife they want; letting constituents pollute streams or air; or writing a technical report that gives granting agencies only the answers they want to hear, any more than a successful business would keep their customers happy by letting them take all of their services or products without paying. Natural resources professionals would ultimately lose the very natural resources they are trying to conserve, and companies would go out of business if they did these things. If you are a researcher, "keeping the customers happy" means providing constituents with accurate, complete information in a timely manner; if you are a resource manager, it means protecting and managing natural resources to the best of your ability.

How does one provide excellent customer service? First, believe in what you are doing. Zig Ziglar, famous salesman and author, claims the depth of your belief is far more important than fancy oratory to sell your product.[2]

Positive projection, absolute conviction, and enthusiasm about your product or service are crucial in selling effectively and providing good customer service. On occasion I have had students who do not enjoy sportfishing and feel that there is no place for exotic fish, which make up the majority of sportfisheries in the western United States. These students typically do not make good sportfisheries biologists, simply because their heart is not in their job, and they have a very hard time selling their ideas to other biologists, their supervisors, and their other customers, the sport anglers. Those biologists who truly believe in their work or their product both consciously and unconsciously project their belief to others, make the most effective "salespeople" of their products and ideas, and provide excellent customer service. This point is important, because you stand to be much more of a success in your job if you can enthusiastically support it. This also applies to employees you hire. Employees who can enthusiastically support your goals and products are critical for the success of your program.

Next, help your customers solve their problems. Zig Ziglar claims you can get everything you want in life if you will just help enough other people get what they want. First learn about your customers, and ask questions to identify their problems. Find out what is needed to solve their problem, and show them how they can solve that problem with your product or service. Perhaps your product or service doesn't exactly fit the bill. Use your imagination to help your prospect get what he wants. In the world of sales, you are solving people's problems. Many people think salespeople and customer service representatives are just master manipulators. However, Ziglar argues that a manipulator may make some sales, but will be a very poor salesperson for the long term. Success comes from a genuine interest in your customer, and an honest desire to help them solve their problem. The Cooperative Fish and Wildlife Research Units program, managed under the U.S. Geological Survey, teaches future conservation professionals in graduate schools around the United States, and provides service and management-related research to state, local, and federal agencies. The program is extremely popular with agency biologists and managers at all levels because it trains future employees, provides service to management agencies, and conducts research to help solve critical management-related problems. This is the core of customer service. Those researchers who are most effective at obtaining funding are those who have excellent relationships with the agency, university managers, and biologists. They go to their meetings, cookouts, and field excursions, all to find out what their problems are and how to help solve them.

Learning who your most important customers are and giving them priority service can increase your effectiveness. Treat these priority customers like gold! Here, as in chapter 7, we apply Pareto's principle or the 80/20 rule.[3] In this case 20 percent of your customers provide you with 80 percent of the benefits of good customer service. These 20 percent of the customers are the ones you work hardest to please. For the Arizona Game and Fish Department, license holders fund a huge part of the department's finances and are therefore a priority customer. In my job, my supervisors expect me to work effectively with the cooperating state and federal agencies for our program. Therefore, my priority customers are the Arizona Game and Fish Department, the U.S. Fish and Wildlife Service, the University of Arizona, the U.S. Geological Survey, and the Wildlife Management Institute. I will provide service and additional information to those who ask for it, such as private consultants; however, the five customers listed above will always get top priority. Learn about these most important customers and tailor your services to them.

It costs far less to keep a loyal customer than to get a new one. Therefore, to keep your customers for life, stay closely attuned to their wants and needs by asking them at all times how you are doing and how you can do it better. You might think you are doing very well in your job; however, to be blunt, the only opinion that really counts is that of your customer. Therefore, at all times being aware of your customer's perception and opinion of you is important if you want to advance. One of the secrets to keeping customers is to reward them. Give them an inducement to buy from you, use your products, support your ideas, or fund your grant. What advantage does that customer gain from working with you? What reason does he or she have to work with you? If there is a reward for doing business with you, the customer *will want to* work with you. One way to reward your customer is to underpromise and overdeliver. Blanchard and Bowles call this deliver plus one.[4] If you promise something to the customer on a certain date, give it to them, plus a little extra. If you are late with something, make sure you give the customer some perks to help make up in part for your tardiness. You often will reward them just by working with them. Will involvement with you look good to their boss? Do they get to collaborate on a publication that enhances their reputation, perhaps a manuscript they would not be able to do unless they worked with you? Do you compliment them for all of the excellent help and advice they provide? Customers who have worked closely with me on projects have been real assets. I get good advice from people who know the research needs intimately. They get the opportunity to interact with graduate students, university faculty, and others. It most definitely is a win-win situation.

A customer may not know that they are being rewarded. Therefore, it is up to you to point out to them and remind them subtly how they are rewarded by working with you. Point out the services you are giving them, especially those that are above and beyond the call of duty.

A final point to remember: you are the company to that customer. Many years ago, when I dealt with the people at a rural lake, they were angry because, according to them, the state agency had reneged on a promise to put trout in their lake. What I learned later was that one misinformed biologist had promised them something the agency could not deliver and which he was not in a position to promise. The landowners believed the entire agency treated them poorly because of their interactions with that one biologist. A note to supervisors—hire people who will represent the agency well and are committed to customer service!

GETTING FUNDED

Let's assume that you are trying to obtain a grant, or proposing to add to your existing program. The first part of getting funded is to pay particular attention to the scientific merit of your proposal—the importance of the question you are asking, and the quality of your research that will be used to answer it. High-quality research used to solve a particularly germane question is the critical first step in receiving attention for your proposal. Is your research answering an important question of the granting agency? I know of a scientist who spent his time studying odd things such as why bluegill's heads were shaped the way they were. It is not hard to see why this scientist might have had trouble interesting agencies in research funds. Have you asked the granting agency what research areas need the most attention?

Where Is a Great Place to Learn About Customer Service? Successful Companies!

If you look in the business section of most book stores, you will see numerous books on customer service. Customer service is often the root of a company's success. Consider Nordstrom for instance—the soft-goods retailer that started in the Pacific Northwest. Nordstrom was profiled by Morley Safer on the CBS show *60 Minutes* for its excellent customer service. It rewards employees who go to great lengths to serve the customer, and stories about the company are legendary.[5] For example, a clerk did not have a pair of slacks a customer wanted, so the clerk went across the street, bought them from a competitor, and sold them to the customer at a discount.

Other successful companies demonstrate an obsession with customer service as well. Thomas J. Peters and Robert H. Waterman Jr. describe the customer-obsessed culture of IBM in the 1970s and 1980s.[6] The best salesmen were made assistants to the top officers at IBM. In that position they spent their entire time answering every customer complaint within twenty-four hours. A vice president at IBM wanted his sales staff "to act as if they were on the customer's payroll." Internal and external customer satisfaction at IBM was measured on a monthly basis.

Fabulous customer service has been a cornerstone in the success of many corporations. Many government agencies are realizing how useful good customer service is as well.

Usually the granting agency will have a list of priority topics. Have you asked other scientists and managers outside the agency about your idea? Did they seem to think it was a good one? Have you had your proposal reviewed by other scientists prior to submitting? Have you had your proposal examined by a statistician who can tell you whether it is statistically sound? Can you really complete the project within the time limit and using the money you are asking for? Have you put your proposal in the format they requested? Finally, have you answered all their questions completely? If you can answer yes to all of these questions before submitting your proposal to a granting agency, you stand a much better chance of getting funded. Completing these steps first is the most critical part of developing a well-written, competitive proposal.

The difficulty is that many of the proposals that are received by funding agencies are well written and competitive. Funding agencies typically do not have to decide between a stellar proposal and many poor ones. They usually have several good ones to select from.

Say you have an excellent proposal, one that is well thought out, answers an important question, and is statistically sound. How do you get the edge up on competitors who have equally good proposals?

First of all, if you have been paying attention to customer service, and develop a reputation for solving people's problems, then you should be getting your share of funding, or at least be in a good position to ask for it. Next, using the influence techniques presented in chapter 4 can be especially helpful for giving your proposal a boost.[7]

Using Influence Principles

Cialdini's influence principles can be powerful tools to give your otherwise solid proposal a bit of an edge on the competition. Now let us look specifically at how these principles can be applied.

Social Proof
When you attempt to obtain funding, you stand a better chance of getting the funding if a lot of people support your idea. I ask as many people and agencies as possible to write letters of support that I can include with grant applications. For example, I wrote a proposal that would support research that solved problems in five western states. I asked the biologists and managers from these states to write letters to the granting agency describing how important this information was to them. This resulted in letters supporting the proposal from five western states and seven agencies. The granters were impressed by the level of support that the project received and commented on it. It was funded under three different granting programs.

Authority

I am currently working on a fish sampling program with nationwide signifi-
cance. Dozens of well-known sampling experts are collaborating on this program with me in addition to two coeditors who have edited the best-selling
fisheries texts in North America. I am not nearly as well known as most of the
people on this project; however, the momentum of so many other experts collaborating with me on this project is a powerful selling point to granting agencies. So far, most federal agencies that work with fish have funded this project.

Rarity

Hopefully, if you pick a topic of extreme interest to the conservation management community you will already have addressed this one. Much conservation research deals with protecting a rare organism from extinction or
preserving a rare opportunity to enjoy our natural resources. Make sure
your granting agency knows what will be lost if you do not complete this
research, and what will be lost if you do not do the project *right now*.

Reciprocation

Most projects stand a better chance to be funded if there is "seed" money
available to match the agency contribution. For example, state funds often
require federal agencies to match the funds they will put toward a grant,
while federal agencies often require state funds. If you have funds, volunteer
labor or other "in-kind" services that you will be willing to provide to support a project, you stand a much better chance of getting funded.

Liking and Similarity

Most successful professors establish a rapport with funding agencies and ask
them what they are looking for in a project. Personal contact makes you
known to the person and allows you to find out what they are seeking. Get
to know people in your funding agencies, fellow managers, members of
various constituent groups. Treat them with politeness and respect. You
may never know from where your next fund source is coming. I once sat
down to lunch with a fellow biologist to discuss a particular project on
which he was working; I ended up walking away from the table with
$160,000 to assist in this project, which when I matched it with other
sources turned into a major area of important research. One hour earlier I
would never have seen it coming.

Commitment and Consistency

If the granting agency has a history of supporting research like yours, you have a better chance of getting a grant. Make sure that in your application you point out how your work will be consistent with the goals of the agency and is similar to other work they have funded. You can find out what the agency goals are by talking with the funding officers, carefully reading the call for proposals, and familiarizing yourself with projects they have funded in the past.

Project Wrap-Up and the Importance of Being Timely

I know a water quality chemist who was in charge of distributing grants worth hundreds of thousands of dollars. He seemed to enjoy his position of influence over who received grants and who did not. He showed me a shelf in his office that contained row after row of reports. He commented at length about the different reports, but did not discuss the technical merit of the reports or the skill of the authors. Rather, he was most impressed by those reports that were submitted on time. Timeliness was incredibly important to him. Since then, I have seen numerous grant officers comment on the importance of timeliness. If you are working in the research arena, submitting reports on time is not always possible due to factors beyond your control. However, whenever possible, try to factor additional time into your study proposals to allow for unforeseen consequences, and do everything in your power not to be late.

The Asian Tapeworm: An Example of Obtaining Project Funding

I will now discuss an example where we used all the above principles to obtain funding for a project investigating the effects of the parasite Asian tapeworm on rare desert fishes. As discussed in chapter 2, the southwestern deserts contain a small variety of fish species found nowhere else on the planet. These fish are unique, many with unusual adaptations; however, they are ill-suited for competing with introduced nonnative aquatic species. The Asian tapeworm, a pathogen dangerous to fish, was introduced accidentally from Eurasia to the United States over thirty years ago. Since then, it has appeared in a variety of places, including the desert Southwest. In the 1990s, Asian tapeworm started to appear in rare desert fishes in isolated pools, far from populated areas. We were interested in obtaining funding to investigate the effects of the parasite and determine what could be done about it.

The topic of study was an important one, of interest to many people. Because we wanted to investigate an introduced species that could have serious effects on endangered desert fishes (the rarity principle), agencies were more likely to fund the project. We were able to get some initial funds from a couple of agencies that had very good relationships with us. They knew our research team, and we had always placed a high priority on serving their needs and getting high-quality research done on time for them (the liking and similarity principle). However, the amount of money these agencies supplied did not provide all the grant money we needed, so we had to approach other agencies we had not yet worked with to get the remainder of the project funding. To do this, we obtained numerous letters (social proof) from high-level management and prominent scientists (the authority principle) of agencies already supporting the proposal and included them with our grant application. We also told the agencies that we had already obtained initial or "seed money" for the project and would contribute that to the project if they would be willing to fund the research also (the reciprocation principle). Because the importance of this project, and the use of several techniques to get funded, we were able to attract the amount of resources needed for the study. As the project proceeds, I will be sure to employ the most important step of customer service—my students and I will do the best job we possibly can on this project and ensure the agencies know about it so we can attract funding in the future.

Because I am a research scientist, the example I used above showed how I was able to get funding for research. However, you can use these principles to fund a wide variety of projects to provide huge benefits for our natural resources. You could use these techniques to acquire land and set it aside for parks or wilderness areas, or to get funding to pay for natural resources programs in schools, support exchange programs for scientists from other countries, or clean up toxic wastes here or abroad. You are only limited by your creativity, energy, and desire to do positive things for future generations.

CONCLUSION

A former supervisor of mine from the U.S.G.S. Cooperative Research Units Program, Dr. M. Lynn Haines, captured the essence of customer service and getting funded. Lynn was very popular with the cooperators across the western United States, and I wanted to know her secret for getting so much support from these agencies. She said, "I listen intently to them and try to determine what their problems are. Then I try to figure out how I can help them solve their problems by treating *their* problems like *my* problems." This is customer service.

CHAPTER SUMMARY

- The key to getting funded more than once is providing consistent, excellent customer service.
- Ensure you are providing the highest quality products, whether that is research, management, or other services.
- Believe in what you are doing. You stand to be much more of a success in your job if you enthusiastically support it, and hire people that can enthusiastically support your goals as well.
- Help your customers solve their problems. Learn about your customers, and ask questions to identify their problems. Find out what is needed to solve their problem, and show them how they can solve their problem with your product or service.
- Learning who your most important customers are and giving them priority service can increase your effectiveness.
- It costs far less to keep a loyal customer that to get a new one. Therefore, to keep your customers for life, stay closely attuned to their wants and needs by asking them at all times how you are doing and how you can improve.
- One of the secrets to keeping customers is to reward them. Give them an inducement to buy from you, use your products, support your ideas, or fund your grant. Underpromise and overdeliver.
- Specifically, to get a proposal funded, it must be high-quality research that addresses a particularly germane question of the granting agency.

- Have your proposal reviewed by scientists, managers, statisticians, or other appropriate reviewers to ensure it is of the highest quality and scientific merit; make sure you can complete the project within the time limit and under budget; and ensure you have put your proposal in the format required by the granting agency.
- To get the edge on equally good proposals, pay particular attention to customer service, and develop a reputation for solving people's problems. Cialdini's influence techniques can also be used to get funding agencies to support your proposal.

CHAPTER SIX

How to Negotiate Effectively

Natural resources negotiation has always been highly important to the United States. In 1781, five years after shooting began at Lexington and Concord, a small upstart group of colonies was negotiating the specifics of their independence with their mother country, England. Three issues remained unresolved: the location of the border, the payment of debts to England, and cod fisheries. The issue of cod fisheries, by some accounts, was the most difficult issue to resolve. John Adams was a chief proponent of New England's right to fish cod in the waters of the Grand Banks, Scotian shelf, and the Gulf of St. Lawrence, all located offshore of Britain's loyal colonies in Canada. He wanted to ensure that any agreement reached after the War of Independence would protect these valuable fisheries for New England fishermen. Cod fishing was important to the American economy. Dried cod was traded for molasses in the French Caribbean, and many times the British tried to regulate this trade, causing hard feelings in the colonies, and aiding in the path toward war. Adams doggedly held that the British must allow Americans to continue to fish in these waters. Other Americans, including Benjamin Franklin, thought they could concede on cod fishing rights. Finally in November 1782, the British gave in to the American demands and let them continue to fish offshore of the Canadian colonies. An agreement was signed one year and one month following the British surrender at Yorktown. According to some, securing the right to fish cod delayed the agreement at the end of the war, and thus lengthened the time until America reached true independence.[1]

Natural resources negotiation remains an important skill for the natural resources professional today.[2] Natural resources professionals negotiate every day, often without being aware of it. An employee negotiates salary with her boss. A biologist negotiates with landowners to allow access to a hunting area. An EPA representative negotiates with a power plant to lessen their impact on air and water quality. According to Lawrence Susskind, Paul Levy, and Jennifer Thomas-Larmer of the MIT-Harvard Public Disputes Program, hundreds of environmental agreements are negotiated each week, many of them to settle impending litigation.[3] They state that 90 percent of these negotiations are settled before they reach trial. However, Susskind and his coauthors maintain that environmental professionals often do not see negotiation as an option, because they do not feel that as "regulators" they can negotiate; they may not recognize when they are involved in negotiation; and they often do not have the skills to negotiate effectively.

When people think of "negotiation" they often think of a 1950s-style smoky room, filled with Jimmy Hoffa types holding hard to a position and forcing the other side to back down. The toughest, loudest person who holds fast to his position often wins through sheer, brute strength. This traditional or position-based bargaining is characterized by the following. You determine what you want, and then you offer an inflated opening demand, called a "position," and expect the opposition to provide a very low counteroffer. You then haggle to reach a settlement somewhere in the middle. You care little for the needs of the other side and you use pressure tactics and intimidation so they will cave in and agree to your position. You continually try to downgrade their data and arguments, and concede nothing without getting something first. Being "tough" in this type of negotiation is often advantageous. Fisher, Ury, and Patton[4] bring up the classic case of Adolf Hitler dealing with Neville Chamberlain, the British prime minister before World War II. Hitler made demand after demand for more territory, and

Chamberlain caved in to these demands. He thought that appeasement of Hitler would prevent war. However, "being nice" in this style of positional negotiation was no answer, and Hitler realized that once Chamberlain was willing to give in to some of his demands, he would probably cave in to more. Here neither party won—Chamberlain did not stop Hitler, and Hitler was ultimately crushed in war.

Positional bargaining is commonly used, but it can be inefficient. This style of bargaining can create enemies of people with whom you really need to develop long-term relationships, such as hunting and angling groups, industry, and other agencies. Negotiators using positional bargaining are put in the uncomfortable position of concealing their true feelings when presenting a ridiculously high or low offer. Further, winning your "position" may not meet your actual needs as well as a strategy of ascertaining the underlying interests of yourself and your opponent.

Let us examine the classic case of the orange that Fisher, Ury, and Patton discuss in their book *Getting to Yes*.[5] Two people fight over a single orange. The position of each person is that "I want the orange." It seems they are at a stalemate because there is only one orange and two people. Therefore, using positional bargaining, the tough, loud one can often get the orange from the nice one. The two parties can flip a coin and the winner can get the orange. However, one party will walk away from the agreement dissatisfied.

INTEREST-BASED BARGAINING

A more efficient method of deciding what to do with the orange would be "interest-based" bargaining. Here, the parties try to figure out why each wants the orange. What are their interests? When asked why they want the orange, one says they want the peel to make orange zest. The other states they want the sections of the orange to eat. Therefore, the orange is split to meet the underlying needs of both parties. One party gets the peel; the other gets the sections. Both have met their interests and both are happy. This is "win-win" negotiating.

According to Susskind, Levy, and Thomas-Larmer, interest-based negotiation is characterized by doing the following:[7]

Teddy Roosevelt: Peacenik

A Rough Rider leading the charge up San Juan Hill, a boxer, a tough cowpuncher, a president who sent the "Great White Fleet" around the world in a show of American might. These are the images most have of Teddy Roosevelt, the U.S. president at the turn of the twentieth century. Historian Paul Boller states that Roosevelt was probably the only president who looked upon war "as a good thing in itself." However, it is little remembered that Roosevelt's negotiation skills enabled him to be the first American to win the Nobel Peace Prize. In 1905, Japan and Russia had been fighting a brutal war for about a year, and both sides were being drained both monetarily and in public support for their cause. Russia and Japan's best alternative to a negotiated agreement (i.e., BATNA), which was to continue to fight the war, was deteriorating over time; therefore, negotiation was needed. Roosevelt showed great diplomatic skill, treating both sides equally and helping them reach a negotiated settlement called the Treaty of Portsmouth (named after Portsmouth, New Hampshire, where the talks took place). Roosevelt said of the Nobel Peace Prize, "There is no gift I could appreciate more."[6]

- Analyze and try to improve your BATNA (best alternative to a negotiated agreement). Raise doubts about their BATNA.
- Focus on underlying interests, not positions.
- Invent options for mutual gain.
- Use objective criteria to argue for "the package" you favor.
- Negotiate as if relationships mattered.

We will now examine each of these steps and describe them in detail.

Know and Improve Your BATNA. Cast Doubts on Theirs

BATNA is an initialism often heard when discussing interest-based bargaining. It is the "best alternative to a negotiated agreement." It is what happens when you decide you or the other party does not want to negotiate, or talks break down and you cannot reach agreement. Examples of BATNAs include:

- I will have to go to court to get this settled.
- I will no longer work with you.
- Our country will go to war.
- I will buy this product elsewhere.
- I will not be happy, but I will continue to work/live/etc. under these conditions.

If you, or the other party, cannot do better through negotiation than your BATNA, then there is no reason to negotiate. Often beginning negotiators make the mistake of assuming that agreements always must be negotiated and an agreement reached. However, you should negotiate an agreement only if you can improve your BATNA.

Let us consider an example from the environmental field. Suppose you are a regulator for a county environmental agency. An individual wants to bulldoze the riparian area of a stream and set up a salvage yard for old cars and car parts. The individual has almost no community support for the venture; he is clearly in violation of county and city ordinances if he builds his lot, and you simply do not have time to negotiate with this individual given the backlog of work on your desk and the needs of other constituents. You might decide it is in your best interests not to negotiate with this individual and inform him he will be in violation of county ordinances and subject to stiff fines or possibly incarceration if he tries to go ahead with his plans.

Alternatively, another individual wants to set up a salvage yard for old cars and old car parts. This person has strong support and backing from the community because he will employ many people. He has a history of being friendly to environmental causes in the community. The land he wants to build the yard on contains both riparian habitat and disturbed urban areas. He exhibits willingness to work with you and the other regulators, and you feel your time would be well spent working with this individual. His position would be "I want to build the salvage yard on my land." Your position may be "I want to prevent him from building the salvage yard on that land." His BATNA might be that he takes his chances with the court and starts to build his salvage yard. He feels that he will probably win in court, but he can't be sure. Your BATNA is that you will charge him with a violation; however, given the urban nature of his land and the ambiguity of laws regarding it, his popularity in the community, and his past support of environmental programs, you cannot be sure that your charges will stick. You are reluctant to spend the huge amount of money and time to fight him if you realize there is little chance for you to win. Therefore, you might both feel that you can improve on your BATNA through negotiation.

You are stronger in a negotiation if you know your BATNA well and continually examine it, because BATNAs frequently change. In the salvage yard example above, the first step of both the salvage yard operator and the government regulator would probably be to know the laws regarding building a salvage yard in this particular area. Each party might also want to know a bit about their opponent. If the salvage yard operator walks away from negotiations, is he or she likely to be prosecuted based on the past history of the regulator and the agency? What is the past history of the salvage yard operator in following environmental regulations? Has he been involved in other cities, and what was his environmental record there? How many people is the salvage yard operator likely to employ? What is the level of support from the community for the salvage yard? Will your boss and the upper-level management of the agency back you up if you decide to prosecute any violations?

BANTAs can change throughout the negotiations. Say the salvage yard operator enjoys community support, but a news story comes out describing how he arbitrarily fired many workers in another city at one of his operations. His community support has now been downgraded, and his BATNA is less strong. There is now a greater need for him to reach a negotiated settlement, and you as the regulator can use this to your advantage. However, if a subsequent news story states that the workers were fired because they were found to be stealing from the company, and the salvage yard operator wins an award in another city because of his excellent work

Knowledge of BATNAs— The Power of the South

During much of the early and mid-twentieth century, senators from southern states held an iron-clad lock on the Senate, holding a large proportion of the committee chairmanships. Because of the relatively recent Civil War, senators knew that it was unlikely that a southerner would be elected president, so the highest they could aspire to was a seat as a U.S. senator. Therefore senatorial seats attracted some of the best and brightest the South had to offer. According to Robert Caro, Pulitzer Prize–winning biographer of Lyndon Johnson, these southern senators studied the Senate's rules and precedents with the concentration of men that would be living by them for the rest of their lives.[8] Knowledge of the Senate's rules, what they could and could not do in the absence of negotiation (i.e., their BATNAs) allowed them to get their way, much of the time over non-southerners or liberal members of the Senate who were not as aware of their own BATNAs. The result was that often important legislation, especially that in the civil rights arena, was blocked for decades.

in environmental protection, the salvage operator's BATNA has improved again, and now the salvage operator is in a more powerful position.

Raising doubts about the other side's BATNA is an important part of the negotiations. If the regulator is familiar with the laws in the city, and finds it is very likely that the salvage yard operator will be prosecuted, he or she can relay this to the operator, thus casting doubts about his walking away from the negotiating table and getting a favorable outcome. If the salvage yard operator has a strong previous environmental record, and he has won previous similar suits in other cities, he can improve his situation by relating this to the regulator.

Vietnam Peace Negotiations—
An Example of a Changing BATNA

BATNAs are not static. What a side has to fall back on in the absence of negotiation can frequently change. The end of the U.S.-Vietnam War was a vivid example of each side either taking advantage of or trying to degrade the other side's poor BATNA. The North Vietnamese and U.S. delegations met in Paris to try to negotiate an end to the Vietnam War. At the same time President Richard Nixon was pursuing better relations with the Soviet Union and China, and North Vietnam feared it would become increasingly isolated.

In October 1972, it looked as if a negotiated settlement would be reached, but talks broke down. To bring the North Vietnamese back to the negotiation table, in December 1972, Nixon initiated a bombing campaign of North Vietnam called the Christmas Bombing. The Christmas Bombing was the most concentrated bombing in world history and was directed at industrial and military targets in the North. Over 12 days, 25 percent of North Vietnam's oil reserves and 80 percent of its electrical capacity were destroyed. Many sources also state Nixon might have been trying to convince the North Vietnamese he was a "madman" and would stop at nothing, including use of nuclear weapons, to win the war. The North Vietnamese, faced with intensified bombing, an unknown nuclear threat, and continued destruction if they did not reach a negotiated settlement, resumed peace talks in Paris on January 8, 1973. An accord was reached swiftly that closely resembled what had been agreed to back in October of the previous year.

Fast forward to March 1975. Richard Nixon had resigned the presidency the previous summer, and the American public was sick of the Nixon White House and all things associated with it, including Vietnam. North Vietnam knew their BATNA had changed. If they went to war and did not negotiate, they probably

(continued)

Vietnam Peace Negotiations (continued)

did not face massive bombing or potential isolation from their communist partners. The Americans would probably not attack them. Therefore, in March 1975, North Vietnam began an invasion of the South. The United States did not re-enter the conflict. By early April the North Vietnamese had captured much of the country and on April 30 their tanks rolled into Saigon. The war was over.[9]

Focus on Underlying Interests, Not Positions

The position of the regulator in the above example is to prevent the salvage yard from going in. The position of the salvage operator is to build the salvage yard and operate it on his land. These two positions are diametrically opposed. However, although the positions cannot be reconciled, you can often effectively meet the underlying interests, and both parties can leave with an agreement that satisfies both sides. For example, the regulator discovers that the underlying interests of the salvage operator are making money, continuing in the salvage business started by his father, and protecting the environment for his grandchildren. The regulator's interests are protecting the environment, working out a good agreement to impress his or her boss, and maintaining good relations with the community.

How do you identify your interests and that of the other party? Ask yourself why they are taking a particular position or why they are not accepting your position. For example, why does the operator want to build the salvage yard there? To make money. Recognizing this interest gives you both power to reach an agreement. You can both brainstorm options that will allow him to make money without harming the creek. You also have a mutual interest in protecting the environment. Together you can brainstorm options that will allow him to continue in the salvage business and still protect the environment for his children and grandchildren.

Usually interests are the most basic, underlying human needs. As a state biologist, I gave a talk at a public meeting. A woman in the crowd stood up and said, "What is going on about sediment deposition in the creek below my house? I have an associate's degree in fisheries, and I will not allow you agency types to walk over these people!" Our presentation had absolutely nothing to do with either the creek below her house or sediment deposition. We were not in any way trying to "walk over" anyone; in fact, we were having the public meeting to solicit public input about a different project. Here, we felt the woman cared less about sediment loads in the creek than trying to meet the basic human need of being seen as important by her neighbors. By standing up to the agency "bureaucrats" in a public meeting, she may have felt that the neighbors would consider her their "defender." Assuming this, I complimented her in front of her neighbors, saying, "You should be commended for feeling so strongly about the creek in your neighborhood and for standing up for your neighbors!" In fact, after the meeting, she came up to me, acted as if there were no hard feelings, and asked for a job! The regulator in the salvage yard example might do well to try to identify the salvage operator's basic needs. Does he have a large ego that needs to be stroked? Perhaps, if the operator decides to work with the agencies, the regulator can arrange as part of the agreement a newspaper story praising the operator's environmentally responsible work. The regulator can point out that by destroying the creek, the operator may anger many people and destroy his legacy in that town. Perhaps making money is related to a safety need of the operator. If the regulator can point out other ways to make money or a way to site the salvage yard in an area away from the creek that is equally profitable, the safety needs of the operator might be met.

Each party to a negotiation may not want the other to know what final position or "bottom line" they will accept. However, it is essential that both parties discuss their interests. A frank discussion of the interests of both sides allows all to explore how interests can be met to help reach agreement.

When there are many parties to an agreement, positional-based bargaining becomes even more difficult. Say there are many more parties to the salvage yard negotiation, including the City Environmental Regulatory Agency, the County Environmental Regulatory Agency, the State Fish and Wildlife Agency, the Chamber of Commerce, the federal EPA, the Salvage Yard Operator, and a local community group, "Friends of Smith Creek." Each may have a separate position. The city and Friends of Smith Creek may not want the salvage operator to build there at all; the county's position is that the operator must build fifty meters back from the stream; and the State Fish and Wildlife Agency position is that the salvage yard should be built away from the woods on the side of the property. Trying to reconcile these various positions is almost impossible. Therefore, when many groups are involved, interest-based bargaining becomes even more important.

Invent Options for Mutual Gain

Once the interests of the various parties to the negotiation are identified, the parties can brainstorm options for mutual gain. At this stage, all sides should be working on the same side of the table, attacking the problem together. Sometimes a facilitator or third party is needed to work with the parties to keep emotions neutral. Ideas that meet interests of both sides can be written on a flip chart in front of the group. No one is allowed to comment on the ideas at this time and parties are not required to accept any suggestion, even their own. The purpose of brainstorming is to encourage free thought. Broad thinking "outside the box" expands the options available for everyone and provides the best chance to reach a negotiated agreement. Remember there is no "fixed pie" that needs to be split. The "pie" can be expanded to include a wide variety of options that will make both parties satisfied with what they get. Once all ideas that might help solve the interests of the parties are put on the flip chart, then the parties can discuss which are most feasible.

As a reminder, the interests of the salvage yard operator are to make money, to continue in the salvage business, and to protect the environment for his kids. Your interests, as a regulator, are to protect the environment, please your boss, and keep good relations with the public. Below is a brainstormed list of ideas that might meet the underlying interests of both parties. As you can see, some ideas are feasible and some are not, but the point of brainstorming is to get as many ideas on paper as possible so these can then be narrowed down to a potential settlement.

- The salvage yard can be built on other property that the operator owns.
- This salvage yard is not built and the operator tries another business here, such as sales of used parts. The salvage yards that are already on other properties are used to carry on the family business.

- The salvage yard can be built a specified distance back from the creek. If he does this, the regulator and chamber of commerce will recognize him as an environmental "steward" in the community.
- The salvage operator can spend money he would have spent on court costs to improve the riparian habitat of the creek.
- The regulator could allow the salvage operator to bend the rules in this circumstance but start with the next permittee to heavily enforce the rules about riparian areas.
- The salvage operator would build a fence around his operation that would keep materials from flowing into the creek.
- The regulator would not stop the salvage operator from building anywhere he wanted, but the salvage operator would be responsible for making sure his operation would not affect the creek. The regulator would allow technicians from the regulating agency to monitor the creek below his land to check for chemical spills or other pollution entering the creek.

Use Objective Criteria to Argue for "the Package" You Favor

Next, use objective criteria to evaluate the ideas and narrow them to a few feasible items. Identifying the laws of the area, relying on previous research about the topic, and soliciting the advice of experts you both trust are all ways of using objective criteria to identify potential solutions. For example, the regulator knows that county ordinances will not allow him to "bend the rules" for this permittee and enforce regulations on future businesses that build along the creek. The laws also state that the operations have to be built fifty meters back from the creek. However, research by the regulator shows that salvage yards that are built one hundred meters back from streams in similar areas have much less effect on stream systems. Research also shows that fences are not effective in keeping waste from entering the creek in this type of operation. You have checked with your supervisors, and they are willing to support you if you decide to enforce the regulations through legal channels. The salvage operator presents a market analysis of the areas and finds that given its industrial character, a salvage operation would have the greatest chance of making money on this site.

After considerable discussion, you arrive at an acceptable package for both parties. The salvage owner will place his yard one hundred meters away from the stream bank and will fund a stream bank revegetation project next to his property. He determines that siting the yard fifty meters closer to the creek, as is allowed by law, will not make him much more money and will create hard feelings within the community. Therefore he agrees to the one hundred meter setback. He will also allow technicians from your agency to monitor his land in the future to determine if his operation is polluting the stream and agrees to take steps to protect the stream if they find problems. You as regulator agree to work with him to ensure he has the necessary permits to carry out his operation. Furthermore, you will make certain he receives publicity and awards from your agency for being an excellent environmental steward. Because he lives in this community, and the community is environmentally minded, this will help him earn money in his salvage operation and allow him to leave a positive legacy.

Negotiate as If Relationships Mattered

If you are a government official, businessperson, biologist, or anyone else involved in a negotiation, chances are that you will be living in the community a long time and will be working for a long time with those with whom you negotiate. Therefore, it is important to keep good relations with the other parties, even while negotiating with them. Here is where you can use your verbal judo and influence techniques (see—it all fits together!).

Use of Objective Criteria— A Personal Example

I was once part of a citizens group, formed to stop a developer from building houses on a ridge top in a mountain range close to our home. Those members of the citizen's group that were really able to halt the building and bring the developer to the negotiation table were not those who got angry with him or appealed to his sense of civic duty. Rather, it was those who knew the laws—what he could and could not do. They used this information, objective criteria if you will, to delay the project and obtain concessions. Having objective criteria was the most powerful tool in this situation.

If someone gets angry at you, inquire what is wrong, and try to understand where he or she is coming from. Next, use agreement in some form (but do not agree so much you give away your interests), strokes, or empathy. Then diplomatically state your point of view. Because of the liking and similarity principle of influence,[10] you are much more likely to get your way in a negotiation by negotiating hard but keeping the relationship with the other parties on a level as friendly and businesslike as possible. As the old cliché goes, you can certainly attract more flies with honey than vinegar!

TWO EXAMPLES OF REAL-LIFE NEGOTIATIONS: THE RESERVE MINING COMPANY AND SNOQUALMIE DAM

Sometimes negotiation suits the parties, and sometimes it does not. Scott Mernitz discusses two case histories in environmental mediation that had very different results.[11] The Reserve Mining Company operated a taconite plant in Silver Bay, Minnesota, along the northern shore of Lake Superior.[12] Since the late 1940s, Reserve was involved in hearings, studies, and conferences about the impact of its release of tailings waste into Lake Superior. The waste contained asbestos-like particles that were alleged to cause cancer. Mernitz contends that there was no impetus for Reserve to negotiate and do anything about the waste until they were forced by court order. They had few shared interests with the environmental groups trying to stop the pollution. They were making $60,000 a day and approximately 3,000 people, or 90 percent of the population of Silver Bay, was working for Reserve in the 1970s. Speculation was that Reserve had plans to close, and they did not want to do anything that might have curtailed their profit making at the time. Consequently, Reserve took a hard line to continue their waste dumping into the lake.[13] According to Judge Miles Lord, when asked if they couldn't find a way to stop dumping tailings into the water, creating dust and polluting the air, Reserve Chairman C. William Verity stated, "We don't have to. We won't." Following this comment, Judge Lord shut down Reserve, but a federal appeals court allowed Reserve to reopen the plant until it could develop an alternative to dumping their waste in the water. In the early 1980s the company started to pump their tailings on land, away from Lake Superior. However, Reserve went bankrupt in the mid-1980s when demand for U.S. taconite and steel declined.

Environmental mediation proceeded very differently during the planning for flood control on the Snoqualmie River in the state of Washington.[14] A major flood hit the Snoqualmie Valley in 1959, prompting the Army Corps of Engineers to consider large flood control dams on the river's middle fork. Environmentalists were concerned that construction of these dams would block access to the Alpine Lakes Wilderness, destroy a section of river popular to whitewater boaters, and encourage development downstream. The governor vetoed the plan to build dams in 1970 and 1973, but mediation was started in 1973 to try to resolve the issue. Unlike the Reserve Mining case, the Snoqualmie Dam controversy was one where the various groups had joint interests. The environmentalists wanted environmental protection, but realized that by blocking a method of flood control, they could be blamed if another large flood hit the region. Farmers and homeowners in the area wanted flood control, but did not want extensive development, and wanted to preserve the rural character of the area.

The final result of mediation was a settlement suitable for all parties. A dam would be built on the north fork of the Snoqualmie, not the middle fork, the river that was of greatest concern to the environmentalists. Raising the spillway on the preexisting Tolt River Dam and building a group of levies along the middle fork would provide additional flood protection. Undeveloped areas along the middle fork would be kept rural by purchasing easements and restricting development. The negotiation was a success.

However, this case had an ironic twist. After the textbook success of this interest-based negotiation, with both parties in agreement, the dam construction did not proceed as planned, but this was not due to any problems in the agreement. Later studies indicated that the site chosen for the north fork dam was not geologically sound. If the foundation of the site had been solid, the construction of this facility would have proceeded with all parties in agreement.

CONCLUSION

Within a few days of reading this chapter you will probably have to negotiate. Perhaps you will have to work out an agreement with your daughter to clean her room or you may have to negotiate a major land agreement among agencies. The negotiation may be a protracted process, taking days or weeks, or it may be over in a few minutes. Try the principles outlined in this chapter and see if you are happy with the outcome. Using interest-based negotiation, you stand a good chance of getting results that are acceptable to all parties while preserving good relationships.

CHAPTER SUMMARY

- Two typical ways that people negotiate are position-based bargaining and interest-based bargaining.
- Position-based bargaining occurs when the two parties stake out a position and argue the merits of that position. Usually the more forceful person wins, or there is an agreement somewhere in the middle between the two positions. This type of bargaining is typically inefficient and often leads to an agreement that does not satisfy one or both parties.
- Interest-based bargaining occurs when the parties focus on the interests underlying their positions to see if an agreement can be made to satisfy both parties. Interest-based bargaining typically consists of focusing not on positions but underlying issues, inventing options for mutual gain, using objective criteria to argue for the package you favor, and negotiating in a way to preserve relationships with the other parties. This type of bargaining is typically much more efficient than position-based bargaining.

CHAPTER SEVEN

How to Manage Yourself

One of the best skills people can have is the ability to manage themselves. I have seen some individuals try to apply the techniques discussed in this book, but fail because of personality quirks, self-doubt, or the inability to organize.

By having the raw materials to participate successfully in the public arena, one can significantly improve his or her chance for success. Managing your time, organizing yourself, and managing your stress are the first steps in enabling you to grow in other ways, such as learning new scientific, managerial, or communication skills. Here we will examine some basic concepts for managing yourself, which will allow you to make best use of the people skills discussed in the rest of this book.

TIME MANAGEMENT

Time is one of our most valuable commodities. This can be especially true for the scientist or conservation professional who is expected to work with the public, conduct studies, manage a budget, acquire funding, and manage personnel. I know a biologist from a large western state who is the only non-game fish biologist for the southern half of the entire state. Being a biologist over such an area is akin to being the U.S. marshal for the entire New Mexico Territory in the Old West. It is a huge, almost impossible, job, and to do it successfully you have to be an expert at managing your time. You may be faced with a similar position in wildlife management, water or air quality regulation, or environmental education—an overwhelming amount of work to handle in a limited period of time. Therefore, we will discuss a few ways to manage your time. Several good books have been written about time management, and I have included a few in the references for this chapter.[1]

Think about it. You have as much time to get things done in your life as Thomas Edison, Margaret Thatcher, Albert Einstein, George Washington, and John Muir did. It is just a matter of your talents and your ability to put those talents to use on things that are important.

One of my friends read that Newton came up with the concept of gravity and invented calculus in one summer. Now there was a time manager! We secretly hated Newton for it. I could not seem to finish writing a few of my dissertation chapters during the same amount of time. This was my typical morning as a graduate student: 8:00 a.m., go get a cup of coffee with my friends; 9:00 a.m., start editing my report; 10:00 a.m., before finishing the edits, remember that I needed to order some supplies; 11:00 a.m., before ordering all the supplies, remember that I had to make some phone calls—make the easy ones first; 12:00 noon, before the phone calls dealing with important subjects were concluded, remember that I had to read some materials for class the next day. By 12:00 noon, I had skipped from one task to another and had finished nothing. Furthermore, many of these tasks were not really important. When talking with others, I discovered that this kind of time mismanagement was common.

Because I wanted to change this behavior, I trained myself in time management. I learned the key to managing one's time is prioritizing tasks and carrying out the prioritized task as far as possible before going on to the next one. How does one decide what things are really important? In 1906, Italian economist Vilfredo Pareto created a mathematical formula that showed 20 percent of the people in his country had 80 percent of the wealth. Dr. Joseph Juran, management researcher working in the United States in the mid-twentieth century, applied the rule to a wide variety of things.[2] The rule, known as either the 80/20 rule or inaccurately Pareto's principle, says that a few (20 percent) are vital and many (80 percent) are trivial. Applied to time management, it means that 20 percent of the effort gives 80 percent of the results, while 80 percent of unfocussed effort generates only 20 percent of the results. For example, 20 percent of your writing effort results in 80 percent of the text. The next 80 percent of your effort is spent on small changes that do not seem to make much (20 percent) difference. Of tasks completed during the day, 20 percent are high benefit and give you 80 percent of the results. These might include planning your experiments, calling a difficult but important legislator, or writing a publication. The other 80 percent of the tasks, perhaps things like making sure there are razor-sharp rows of pencils on your desk, making sure you read a two-month-old trade magazine, or talking in the hall for a couple of hours with some of the other biologists, give only 20 percent of the results. Therefore, identifying those 20 percent of important tasks and spending your time on them to get you 80 percent of the payoff is key.[3]

Alan Lakein, time management guru, wants you to ask a single question: "What is the most important thing I could be doing right now?"[4] And then do it. Many people will not ask this simple question. They spend their lives doing things that are comfortable without high payoff, such as useless paperwork, water-cooler conversations, or cleaning their desk, instead of working on a high-priority task. Make no mistake, sometimes visiting in the hall with people or cleaning your desk *is* the most important thing you could be doing. Especially if your social standing in the agency is being questioned or if your desk is a mess and it is impossible to find important papers. However, if you are concentrating on low-payoff tasks just because they seem easy, then you might need to reevaluate what you are doing with your time.

How do you implement a time management program? The key is writing everything down and then prioritizing it. The steps reported by many time management books are similar.

Bureaucrats Can Manage Time to Do Important Things, Too!

Claim you don't have time to do important things because of the bureaucracy and red tape? Check out these bureaucrats! The first bureaucrat graduated from forestry school and got a job with the U.S. Forest Service in Arizona. He was assigned to the post of supervisor of the Carson National Forest in New Mexico and later accepted a transfer to the U.S. Forest Products Laboratory in Madison, Wisconsin, where he was associate director. During his years with the forest service he was a back-country ranger, a game and fish ranger, and a supervisor.[5] The second was an aquatic biologist for the U.S. Fish and Wildlife Service. She wrote radio scripts and supplemented her income writing feature articles on natural history for the *Baltimore Sun*. She began a fifteen-year career in the federal service as a scientist and editor, and later rose to become editor-in-chief of all publications for the U.S. Fish and Wildlife Service.[6] Just a couple of paper-pushing bureaucrats, correct? Wrong! These were arguably two of the greatest environmental writers and thinkers of the twentieth century. The first was Aldo Leopold, author of *A Sand County Almanac*, and the second was Rachel Carson, author of *Silent Spring*.

The Master List

First, buy a notebook or handheld computer and record in it every idea, assignment, call, project, task, or errand—large or small, minor or important—as it arises. This is your master list. Keep your master list open and next to your desk. Then, if you think of something else to do while you are working, simply write it down on your master list *instead* of just starting a new task without completing the old one.[7] Below is what a page out of my master list might look like:

- Turn in the travel voucher for my trip to New York.
- Finish editing Andy's paper.
- Analyze data and write up a report on surveying methods for high mountain lake amphibians.
- Buy a birthday present for my daughter.
- Read the Jones publication on distribution-free statistical methods.
- Make corrections to northern pike article.
- Call Sue about employee space issues.

If you have a large project, divide it into smaller, more manageable components on your master list. Tackling a large project without breaking it up can be daunting to most people. Psychologist David Burns gives an example of how people eat to describe this phenomenon.[8] Most of us have no trouble eating. In fact, we enjoy it! However, suppose the entire amount we consume in our lifetime was put in a gymnasium and we were told to eat it. We would be dumbfounded by the task, and we would probably not feel the least bit hungry. Because we are able to split the task of eating up into small pieces, it is not only doable, it is enjoyable. Approach your large tasks in a similar manner. Divide them into manageable components and suddenly they do not seem so daunting—they may even be enjoyable. Set intermediate goals that can be met, and reward yourself when you meet them. Let us take one of the tasks from the list above and break it down so it is manageable.

- Analyze data and write up a report on surveying methods for high mountain lake amphibians.

When you look at this task as written, you might think, "Wow, I'll never get this done. I think I will just answer some phone calls or play on the computer." However, look at the same task when you divide it into subtasks below. Suddenly the subtasks seem more doable.

- Write introduction section
- Write the methods section
- Enter data in spreadsheet program
- Analyze data with analysis of variance
- Put the results in tables
- Write the results section

- Write the discussion section
- Edit the report
- Incorporate comments and provide final report

Review of Your Master List

Next, review your master list daily. Continue to subdivide large tasks. Eliminate the tasks you can. Ask yourself, what is the worst thing that will happen to me if I do not do this? If nothing much, then that task probably has a lower priority than some of the others. There may be an immediate drawback to not completing some tasks—for example, failing to put out a fire or not finishing the report your boss wants today. Other tasks may be equally important for your future, such as getting your degree or learning Spanish if you hope to work in Mexico. Keep these long-term but equally high-payoff tasks high on your priority list as well.

Some tasks you can delegate to others. If there are things you can delegate, by all means do so. Although it can be tempting to hang on to tasks with which you are comfortable, you will get more done if you can entrust as many tasks as possible to your subordinates and concentrate on only the highest payoff, most challenging tasks.

When you are reviewing your master list, start a daily list and also have a calendar. Some tasks you might defer to a calendar to be conducted on a specific date. Other more immediate tasks you should put on your daily list. Once tasks are delegated, eliminated, deferred on a calendar, or added to the daily list, cross them off your master list.

Let us examine our list again and show how we might review it.

- Turn in the travel voucher for my trip to New York—*delegate to secretary*
- Finish editing Andy's paper—*important—move to daily list*
- Amphibian report: Write introduction section—*The amphibian report is due June 1. It will take approximately three months to complete plus one month additional time just in case it is needed. Therefore, defer all amphibian report tasks to calendar. Put note in calendar to start the amphibian report tasks on February 1*
- Amphibian report: Write the methods section—*calendar*
- Amphibian report: Enter data in spreadsheet program—*calendar*

- Amphibian report: Analyze data with analysis of variance—*calendar*
- Amphibian report: Put the results in tables—*calendar*
- Amphibian report: Write the results section—*calendar*
- Amphibian report: Write the discussion section—*calendar*
- Amphibian report: Edit the report—*calendar*
- Amphibian report: Incorporate comments and provide final report—*calendar*
- Buy a birthday present for my daughter—*important—her birthday is tomorrow—move to daily list*
- Read the Jones publication on distribution-free statistical methods—*not important right now—leave on master list*
- Make corrections to northern pike article—*important—Jon is waiting to publish this—move to daily list*
- Call Sue about employee space issues—*important—employees are extremely cramped and this is affecting their productivity and happiness—move to daily list*

The Daily List

From the review of your master list above, you have deferred (amphibian report activities) or delegated (travel voucher) some items off the master list, either to a calendar or to someone's inbox (the secretary). Furthermore, you have identified high-priority items that are then incorporated onto a daily list. Now we will discuss the daily list in more depth. A daily list is a group of items you want to complete during that day, and can be put on a computer, in a notebook, or on a calendar—somewhere separate from the master list. The daily list should include activities to help you meet your most important goals, and get you 80 percent of the results from 20 percent of the effort. Your daily list will usually have about eight to ten items, things you can reasonably expect to get done in one day. Here is a short daily list that was developed from the master list example above.

- Buy a birthday present for my daughter
- Make corrections to northern pike article
- Finish editing Andy's paper
- Call Sue about employee space issues

Now, how do you decide which of these tasks to do first during the day? The high-payoff, high-priority tasks are those you should have put on your daily list. Now rank tasks on your daily list with an A, B, or C, with A having the greatest priority and C being less important. Do the As first, during your prime time.

Most people have prime time. Prime time is the time they work most effectively during the day. My prime time is right when I get into work in the morning. I have a lull in energy around noon, and I get more energy in the late afternoon. Knowing this, I schedule my most challenging, highest-priority tasks during the time I first get to work and in the late afternoon. During the noon hour I do lower energy/priority tasks such as exercising, phone calls, and general paperwork.

- Buy a birthday present for my daughter—*A*—*but prime thinking time not needed for this. Put first thing in afternoon*
- Make corrections to northern pike article—*A*—*put during prime time*
- Finish editing Andy's paper—*A*—*put during prime time*
- Call Sue about employee space issues—*B*—*important, but because employees are gone for two days, could be done tomorrow as well with no harm done. Put at end of day*

When I get into the office in the morning, I will edit and make corrections to the two papers in my prime time. I will get my daughter's birthday present early in the afternoon, and if I have time left over, I will call Sue. This way I can ensure the most important tasks are done first.

Avoiding the Timewasters

Even if you organize your tasks in this manner, you still may have a problem with procrastination, especially on tough tasks that require a lot of effort. Procrastination can be an annoying habit. Usually the highest-payoff tasks are those that are most difficult. Therefore, we tend to procrastinate on them. How can you whip procrastination? First, get started on something *before* you feel like doing it.[9] Often it takes our "pump being primed" to get going on those high-payoff tasks. If we wait until we feel like doing something, we often never get started. If you start on the task, even a very difficult one, it is human nature for you to start to enjoy the task and want to continue because you are "developing a habit." If you still have difficulty starting a task, tell yourself you will only do it for a set period of time, say a half hour. It is surprising how many people, once they get going, actually want to continue following the half-hour startup time. Furthermore, you might promise to reward yourself in some way; for example, take yourself to the movies or a nice dinner when you finish a particularly arduous task.

Here's a Novel Way to Avoid Procrastination!

Procrastination plagues many people, including the famous. Victor Hugo, writer of *The Hunchback of Notre Dame* and *Les Misérables*, would sometimes procrastinate. To solve this problem, he would give a servant all of his clothes, sit in a bare room, and instruct the servant not to return for several hours. Left to his own naked self, with nothing but a pen and paper, he would be forced to sit and write.[10]

Managing interruptions is another important aspect to time management. Some people manage interruptions by posting "do not disturb" office hours, keeping distractions around the office to a minimum, and allowing the phone to record messages. Others might go to a quiet place where they will not be bothered, such as a library. Keep asking yourself "Lakein's Question"—"What is the most important thing I could be doing right now?"

How do you say "goodbye" nicely to well-meaning people who come into your office and interrupt you? One way is to discourage them from sitting down! Have something on your chairs, such as a pencil, books, or paperwork. When there is a brief lull in the conversation, say "Thanks so much for stopping by, but I better let you get back to work." At the same time you can stand up and move toward the door with your hand out to shake theirs on the way out. This works for 99 percent of people. If none of these work, and the person is really being a pest, simply use verbal judo. "I really appreciate you stopping by and visiting. I'm afraid I have to get working on this assignment now or my boss will kill me! Is there another time we could talk?" (Use the last comment only if you would want to continue talking with the person!)

Many people who have five minutes—say before a meeting, waiting in line, between phone calls—will fritter it away, playing solitaire on the computer, talking to someone, or looking through a catalogue. They might say it just is not enough time to get started on something! Don't waste these small bits of time. All kinds of work can be done when you least expect it. I have written entire articles, later published, exclusively during many short snippets of time (five to fifteen minutes). Time management expert Alan Lakein wrote his entire book in the short periods in the morning before his family got up. Often these short snippets of time are the only extra time available during the day.

One of the final bits of advice on time management comes from the famous fish biologist Wayne Hubert. He said that "given the choice between two manuscripts, finish the one that is closest to being done." That advice has worked very well for me on a wide variety of jobs. If you are given the choice between several tasks of equal priority, spending your time to finish the one that is closest to being complete will give you more finished products in the end.

ORGANIZING PAPERWORK AND EQUIPMENT

The government employee worked hard on the memo that detailed some of their findings. They had studied this issue and thought that others should be informed. The memo was sent to upper management, where it died. No action was taken, and the memo was not distributed to others in other agencies that should have seen it. The memo was from an FBI agent and reported that suspicious Middle Eastern men were taking flight lessons in training schools in the Southwest. Furthermore, the memo named Osama bin Laden, suggesting his followers could be using the schools for terrorist operations.[11] One month later, two passenger jets slammed into the World

Trade Center in New York City. Another crashed into the Pentagon, and yet another crashed into a field in Shankstown, Pennsylvania. They were hijacked and flown into their targets by the same Middle Eastern men who had taken the flight lessons reported in the memo.

Mishandling your paperwork may not have the drastic effects of the flight school memo; however, the ability to handle paperwork effectively can dramatically improve your job performance. It can improve your customer service because you will not lose important notes from people. It can cut down on the time you spend looking for key papers. You will not miss deadlines because you misplaced a memo. If you cannot see your desk, waste considerable time finding a paper, or have lost a valuable contract, you might need to reassess and streamline your paper-handling system.

The TRAF System

Organization expert Stephanie Winston states that when you get a piece of paper in the mail, you should handle it only once, acting on it or making a decision about it. Winston claims there are only four and a half ways to treat any piece of paper, and calls this system TRAF: trash, refer to someone else, act, file, and the other one-half, read.[12] If you do this with every piece of mail that comes in, your paperwork will soon be organized. Let's look at the steps individually.

Trash
Winston reports that an executive's best friend is a trash basket—or for the environmentally minded, the recycling bin. Throw away as much of your paperwork as possible. Is it something widely available, such as a company report or an advertising flier? Ask yourself, "What is the worst thing that can possibly happen if I throw this piece of paper away?" If your answer is "nothing much," then you can probably safely discard it.

Refer
If the task in the paper can be handled better by someone else, pass it on. I sometimes get letters asking about the management of birds in Tucson. I am a fisheries biologist. I could try to answer the questions, but I know that they will get a much better answer more quickly if I refer them to professors who

work with birds. Therefore I rapidly pass on the letter to one of these professors. In addition, if your staff can take care of the task, by all means delegate! It makes little economic sense for you to respond to a letter that could just as easily be answered by someone making half your salary.

Act

You have referred or trashed the paperwork you could. Now you take action on the remaining paperwork. Get a person the answer they need, fill out the report for your boss, sign the expense report. If it is something you can do quickly, take care of it immediately. If it is something that takes longer, break it up into pieces and work each piece into your master/daily list procedure discussed in the time management section.

File

Some pieces of paper, such as data on a research project, or a note about an important scientific finding, need no action, but you will want to keep them available. These pieces of paper can be filed. A key to good filing is creating broad categories. If categories are too narrow, it is tough to find things later. For example, if you are a bird biologist you might have a file called mallard ducks. Unless you were a waterfowl or mallard expert, it might not pay to have files labeled mallard duck kidney function, mallard duck reproductive habits, mallard duck feeding habits, et cetera. Having categories that are too small makes papers hard to find later.

Read

The additional one-half of the TRAF system is read. Following reading you can put the paper through the next appropriate TRAF action.

What about emails? Reading and answering emails can take up your entire day. If you have the flexibility, pick one time of the day when you read and answer your emails. The good news is that emails, unlike paper, do not take up much space, and you can safely keep almost all of them in a file on your computer hard drive without the clutter that paper brings. However, you may want to keep paper copies of the most important emails in a file as well. Emails are good records of what you said to others, or can provide a timeline of the history of events. Make sure to back up your emails.

What to Do If Your Workspace Looks Like a Disaster Area

What do you do if you already have huge mounds of paperwork, books, and trash on your desk, and dozens of Post-it notes on your walls? First continue to use the TRAF system for all incoming paperwork. But set aside some time to tackle your desk (or floor!). Weekend days are best so that you are not disturbed. Let's say your desk is completely covered. Pick one quarter of your desk and start on the stack. Return books to the shelves, TRAF all papers, and write information from Post-it notes or other jobs to do in your notebook (your master list). Once that quarter is done, start on the next quarter. Work from the top of the stack down. When you have tackled your desk, you can start on cluttered tables or chairs in your office. You can use the TRAF system to organize your computer files as well.

Organizing Field Equipment

Field equipment can also rapidly get disorganized. If not cared for, equipment can end up broken, spread over a large area, or lost. To organize equipment, first find a secure storage area. Unfortunately field equipment kept in a common area is often pilfered by others. For example, another biologist may need to borrow a tagging gun for a short project, and has every intention of returning it at the end. Frequently they will not return it or you will be unable to find it when you need it. Therefore, keep your field equipment locked up, either in a separate room, or in a locked locker or cage area in a common warehouse. When you organize your field equipment, care for it and store it as a firefighter would. You need to access and load your equipment quickly. Therefore if everything has its place, and is cleaned and stowed there at the end of the field day, you will be able to readily access it the next time you need it.

COPING WITH STRESS, DEPRESSION, OR ANXIETY

You can learn all the time management, communication, personnel, negotiation, and influence skills possible, but if you have problems with stress, depression, anxiety, or other mental health issues, you will be much less effective applying the lessons you learn. Recognizing that mental health issues are real, and can seriously reduce your effectiveness at your job and your relationships with others, is a crucial first step in tackling these problems and moving on. Natural resources professionals commonly have days filled with stress. You will speak in front of large, sometimes hostile audiences. You will have too much work for the time you have available to complete it. You will work with angry constituents, members of other agencies, or coworkers. No one would think anything of your arm being injured if you lifted weights all day. However, somehow the mental equivalent of that example is often ignored. Some people say a person should just "work through it," or else he "is just weak." Using the example of your arm, continuing to ignore the pain and continuing to lift weights would cause further injury. Sometimes tasks can become mentally overwhelming and just as painful. A wise natural resources professional recognizes this fact and does not condemn him- or herself when dealing with challenging mental issues. He or she accepts themselves, recognizes the signs of mental health problems, and tries various things to help these problems get better.

There are serious mental illnesses that are beyond the scope of this book, which I will not discuss here. However, most people have some form of stress, depression, and anxiety in their lives, and learning to recognize them and then heal them is important for professional effectiveness as well as personal satisfaction.

When our job becomes overwhelming, we can experience stress. What is stress? Stress is the physical and emotional reaction you experience as the result of changes or demands in your life. It is important to realize that stress is a part of everyday life. It can be either positive or negative. David Burns gives the example of how you hold your hand in describing bad and good stress.[13] If you hand is held limply, it just sits there, not much good for anything, analogous to a very low amount of stress. If you tightly clench your fist, it is rigid and you cannot get anything done, analogous to a very high-stress situation. Now take your hand and move your fingers. Make it come alive. This is a medium-stress position and is the most productive. People need some stress to function effectively. The key is to not let it become overwhelming.

How can you recognize signs of too much stress? You might have physical symptoms such as headache, backache, stiff neck, rapid breathing, sweaty palms, upset stomach, high pulse rate, or feel jumpy and exhausted. You might have mental symptoms such as irritability and intolerance of minor disturbances, loss of temper frequently, or inability to concentrate.

How can one cope with excess stress? Have ways to express yourself besides work. Relieve stress by exercising or talking with others about the problem. Find another way to express yourself: a hobby, arts, writing, crafts, and relaxation skills.[14] Just getting your mind off your stress by having a life outside of work can be a fabulous way to deal with stress. In his autobiography, Lee Iacocca states that management had to institute a rule at Ford Motor Company that people would have to leave work at 9:00 p.m. Some people literally "dropped dead" at their desks because of overwork and accumulated stress before this rule was enacted.[15]

When dealing with stress, it is important to keep things in perspective. A friend and fellow biologist always had an excellent way of dealing with stress. When discussions on things like fishing regulations and modifying the fishing pamphlet started to get heated, he would say, "Hey, we're not splitting the atom here. Let's keep what we are doing in perspective." He was right! Knowledge of history can be very important. Think of people like the Apollo astronauts who faced life and death continuously on their way to the moon, Dwight Eisenhower on the eve of D-Day who had to make the decision to land during inclement weather, and the fantastic stress Captain James Cook must have endured while exploring the world. Looking at the big picture: Is anything in your life really so bad you cannot handle? Is not giving a good talk at a conference or mishandling a rule change at the department really worth getting upset about when you compare yourself to what others have dealt with?

Find creative solutions to the stressor. Can you solve a problem in a different, less stressful way? Perhaps you have too much work. Can you delegate some of your lower-priority tasks to others? If your bills cannot all be paid this month, can you make arrangements with your creditors to pay two out of three? If you are bickering with a fellow employee, is there a way you can split the tasks up so you can minimize your interactions with each other?

A professor in my department walks off her stress in the nearby Sonoran Desert. She tells me that it is very hard to stay stressed once you are out in the desert, listening to the wind and the sounds of birds. Soon she comes back, relaxed and ready to start again where she left off. Abraham Lincoln

Failed at Something?
Join the Club!

Stress, anxiety, or depression can be triggered by a professional or personal setback. When a setback happens, people might believe they are "failures." However, failure is experienced by everyone. It is not lack of failure, but the fact you keep trying *in spite* of failure, that will ultimately lead to success. Let us examine failures and setbacks of famous naturalists and conservationists who were ultimately fabulously successful.

- Jane Goodall, primate expert, had four students kidnapped from her camp in Tanzania by rebels. Because of unfounded rumors that their kidnapping may have been due to her negligence, it was suggested she leave her professorship at Stanford.[16]
- Gregor Mendel, botanist and genetics pioneer, failed his teaching exams three times and thus could not get a permanent teaching position.[17]
- John J. Audubon, painter and naturalist, lost his sawmill business and a steamboat to unpaid creditors. He was shunned in his hometown, jailed because of his inability to pay further debts, and stabbed a man who threatened him.[18]
- Aldo Leopold, naturalist, environmental writer, and fish and game manager, wanted to be director of conservation for the state of Wisconsin. Despite widespread support for Leopold, the assistant to the governor, a man with no conservation experience, was chosen instead.[19]
- William Beebe, famous underwater explorer, lost his wife to another man.[20]
- Constantine Rafinesque, naturalist and professor at Transylvania University, published a landmark book on fishes of the Ohio River. Included in his book were several fish species that did not exist. Drawings and descriptions of

(continued)

Failed at Something?
(continued)

each species were presented to Rafinesque as a practical joke. One species, called a Devil Jack Diamond Fish, supposedly was between four and ten feet long and weighed four hundred pounds. When it was discovered that these fish were faked, Rafinesque was ridiculed and lost credibility. The man who played the prank on Rafinesque? John J. Audubon.[21]

was asked why he joked much of the time. Lincoln replied, "I laugh so I do not cry." In the deepest days of the Civil War, an overbearingly stressful situation, Lincoln realized that having a sense of humor was extremely important to meet the day's challenges.[22] Try to keep a sense of humor. Surround yourself with upbeat, fun friends, watch comedies, and make humor an important part of your life.

Contrary to what Mississippi blues singer John Lee Hooker sings, one bourbon, one scotch, one beer is not a good way to relieve stress. Drinking, overeating, smoking, and using drugs are not effective at relieving stress and make the problem worse. If you are feeling stressed, avoid these at all costs.

What if you try some of these techniques and nothing seems to work? What if you are "stuck" feeling bad or anxious for weeks or even years? Depression and anxiety are common. Depression is called "the common cold" of metal illness. It is usually characterized by a hopeless mood and loss of enjoyment in activities that used to be enjoyable. Anxiety is characterized by a feeling of fear. A person suffering from anxiety often experiences an inability to concentrate, insomnia, tension-type headaches, and upset stomach. The National Institute of Mental Health estimates that in any one-year period, nineteen million Americans are clinically depressed.[23] Anxiety disorders affect nineteen million people in the United States every year as well.[24]

Depression—
Deadly Even to the Famous

Depression can be deadly, affecting both the very successful and the obscure. Unfortunately, modern help was not available to save one of our greatest naturalists from a horrible death due to depression. Meriwether Lewis, who explored the West with William Clark, was doing a poor job as governor of Louisiana. On a trip back east to explain his lack of success, he attempted suicide twice, first by trying to jump overboard from a steamboat, then by shooting himself. The commander of a fort where Lewis was visiting was so concerned about him that he put him on a twenty-four-hour suicide watch for a week. Lewis convinced the commander that he was feeling better and continued on his travels. Soon after, Lewis rented a room in a cabin in Tennessee. While there, Lewis shot himself twice and cut himself from head to toe with a razor to finish himself off.[25] If you have depression, make sure you get treatment. It could save your life.

Having these diseases can severely alter your effectiveness with colleagues, the public, and supervisors. In addition, these afflictions can make the person having them feel very bad, sometimes to the point of being life-threatening. I know of a scientist who had a very bad case of anxiety disorder. He got along well with his family and peers, but had controversial theories that he thought would make his family, neighbors, and church angry. He was to write a book, but he held up for a protracted time period out of fear of what his family and friends might think. His anxiety got so bad that he complained of heart trouble, vomiting, and insomnia. Things continued to get worse. He had bouts of extreme crying, was afraid of many things, and became a semi-invalid and a hermit while he was still in his thirties. His isolation from people gave him a chance to write his book, which actually did turn out to be fairly controversial. The book was *The Origin of the Species*. The scientist was Charles Darwin.[26]

Anxiety and depression can affect a wide variety of people, in all occupations. Famous people who have suffered from clinical depression include Ernest Hemingway (writer), Abraham Lincoln (president), Mike Wallace (TV journalist), Tennessee Williams (playwright), Buzz Aldrin (astronaut), Kirk Douglas (actor), Laurence Olivier (actor), General George S. Patton (soldier), and Richard Nixon (president).[27] Famous people who have suffered from extreme anxiety include Charles Darwin, Sigmund Freud (psychologist), Barbra Streisand (singer), Nikola Tesla (scientist), and Edvard Munch (painter).[28]

If you have either disease, you may be reluctant to seek help, because you are afraid people might brand you "weak" or "odd." However, seeking treatment may change your life for the better. How do you know if you have depression? Ask yourself the following questions: Do you feel sad much of the time and can't snap out of it? Are you disappointed in yourself, feel like a failure, and sense that there is not much hope of anything getting better in the future? Do you have any physical symptoms, such as loss (or gain) of appetite, irritability, sleep problems, or constant fatigue? Are you not very interested in people or activities you used to enjoy? Can you answer yes to several of these questions and have had these symptoms for more than two weeks?[29] If so, discuss these problems with your doctor as soon as possible. Of course, seek immediate care if you have thoughts of suicide or doing harm to others.

How do you recognize excess anxiety? Some anxiety is natural; however, if you are plagued by persistent fear, worry, or anxiety, if you avoid specific types of situations, or if you demonstrate obsessive-compulsive behavior, you might have a type of anxiety disorder. If you are experiencing some of these symptoms, seek medical help.

Poor mental health can be very costly to a corporation or government entity. According to Johns Hopkins University researchers Alan Langlieb and Jeffrey Kahn, the annual cost of depression and anxiety to U.S. businesses exceeds $145 billion.[30] The good news is that these common mental health problems respond excellently to treatment. If you recognize these diseases and take steps to deal with them, your effectiveness at work and your enjoyment of life will skyrocket. But only if you seek treatment!

CONCLUSION

If you are able to manage yourself well in areas such as time management, organization, and mental health, your profession can be a fabulous adventure. If you need to develop any of these areas further, by all means investigate the excellent references on these topics cited at the end of this book. Once you master these skills, you will have a good foundation with which to employ the verbal judo, negotiation, customer service, and persuasion skills discussed elsewhere in this book. And more importantly, you will have the peace of mind needed for a successful, satisfying career and personal life.

CHAPTER SUMMARY

- Time management, good organization skills, and stress–mental health management are fundamental for the conservation professional to be effective in other areas of the job.
- To manage your time, realize that 20 percent of the tasks provide most results, while 80 percent of the tasks really do not provide much payoff. The key to time management is to use your time to concentrate on the 20 percent high-payoff tasks.
- Write all items that you need to do on a master list. From the master list, enter tasks onto your daily list, delegate tasks, or cross them off as unimportant.
- Free up additional time and avoid interruptions by politely ending conversations, or arranging your office so people cannot sit down and talk easily. Complete highest-priority tasks in your "prime time." Avoid procrastination by starting on a task before you feel like doing it and by breaking large tasks up into small pieces.
- Manage paperwork by handling all pieces of paper once. You can only do four and a half things with each piece of paper. Throw it away if it is replaceable, refer it to someone else if at all possible, act on it, or file it. The other "one half" action is to read it.
- Stress is an everyday part of your work, and too much can affect your ability to succeed. Reduce stress by having other ways to express yourself besides work, finding a creative solution to your stressor, keeping things in perspective, and using your sense of humor.

- Depression and anxiety result from unchecked stress or other factors, such as your body chemistry. These diseases result in billions of dollars of lost work and revenues annually.
- Depression is usually characterized by a hopeless mood and loss of enjoyment in activities that used to be enjoyable. Anxiety is characterized by a feeling of fear. A person suffering from anxiety often experiences an inability to concentrate, insomnia, tension-type headaches, and upset stomach.
- These common mental health afflictions are easily treated. Treatment considerably increases your ability to cope with work and improve your enjoyment of life.

CHAPTER EIGHT

How to Effectively Manage Personnel

The first things you noticed about Tommy were his bright eyes, thick glasses dwarfing his face, and his shock of unruly white hair. Top that off with his wide smile and quick step and you immediately liked the man. I went to Tommy to ask him to edit my job application. It was tough to get people to look it over because it was a boring job, and, since I was only a student, many people simply did not have time. But Tommy did. He spent almost an hour carefully reading it and talking it over with me. Of course Tommy had time for other things, too. Like being elected to the National Academy of Sciences and receiving its prestigious Garner Cottrell Award for Environmental Quality. Or winning the G. Evelyn Hutchinson Medal of the American Society of Limnology and Oceanography for his ground-breaking ecological research. Or being given the Eminent Ecologist Award from the Ecological Society of America. Or being awarded a Resolution of Respect by the Washington State House of Representatives. Tommy won these and numerous other awards for his work. You see, Tommy, also known as W. T. "Tommy" Edmondson, was one of history's greatest freshwater ecologists.[1] And even so, he had time to read my letter.

According to all accounts W. T. Edmondson was an excellent boss.[2] Technicians, whose job tenure is usually measured in months in the ecology profession, worked decades for Tommy Edmondson. In his book, *The Uses of Ecology*, W. T. Edmondson made the argument that technicians should be respected as seasoned researchers and paid accordingly.[3] He made sure to praise the graduate students who had come through his lab, and give them credit for the projects they had conducted. The continuity, motivation, expertise, and high morale of employees—and of course Tommy's brilliant ecological knowledge—combined to provide fundamental breakthroughs in long-term ecological research on lakes and their communities. Tommy could manage employees. And your success as a biologist, scientist, or manager will be closely tied to your ability to manage employees as well.

Effective staff are critical for project success and enhance your reputation, while poor staff can ruin your efforts. On research projects, staff time is usually the most expensive component, and in agencies, salaries are a major portion of the budget. So how do you get good people, retain them, and work well with them? In this chapter we will discuss techniques for hiring and motivating your staff. Some you might already know—others you might find surprising.

HOW TO HIRE GOOD STAFF

Many management specialists consider hiring the single most important part of personnel management. Jeffrey Fox, marketing consultant and business writer, states that a supervisor should "hire slow and fire fast."[4] If you use great care to hire the right person, the number of times you will fire substandard employees will be few.

Given the importance of employees and their expense—and the huge headaches involved in getting rid of someone—it is amazing how many supervisors do not work hard to attract the best applicants for positions. Have you seen intra-agency hires of people who may not want the job, are barely qualified, or are mediocre, yet have considerable seniority? These people are often hired because supervisors do not spend the time needed to find the best person they can. Perhaps they believe that a mediocre person they know is better than an unfamiliar person. Possibly they feel that agency rules will not let them pick the best person for the job, and they depend on the staff of the personnel office to pick their employee. Sometimes supervisors, even ones who are very experienced, do not know what is allowed when hiring people for a job. Fortunately there are a couple of easy ways to dramatically increase the quality of your candidates.

Widen Your Applicant Pool

If you develop a large applicant pool, chances are good it will include someone well suited for the position. Let as many people as possible know about the position. Advertise in trade publications such as *Science* and *Fisheries*, or conservation society websites such as the *American Fisheries Society* job postings. Let colleagues know about an open position through group emails and personal telephone calls. Contact people who are successful elsewhere and ask them if they are interested in a position within your program. Provide a closing date for the position and ask people to respond by then.

An agency supervisor might be restricted to hire someone who is first on an "agency list" of applicants maintained by the human resources department. You can hire capable people from these lists, but often the candidate at the top of the list is not best suited for your position. Therefore, you need to know the regulations so that you can control, as best as you can, who you can pick from the list—or, depending on the job available, if you might choose someone who is not on list. That is why you need to get to know people like Penny Cusick.

Penny Cusick and her staff at the personnel office at the Washington Department of Fish and Wildlife did more to protect natural resources than many biologists. Penny had a professional, pleasant demeanor and an encyclopedic knowledge of the rules and regulations regarding hiring. She and her staff helped managers work within the system and negotiate through the endless hiring rules to pick the best people for jobs. Find someone like Penny in your organization that you can trust and who will give you the straight scoop about how to hire your best choice. Can the applicant you want advance to first on the list, and if so, how does he or she do this? The key is to find out how *you* can control who gets the position.

Call References to Ensure Person Is Skilled *and* Works Well with People

Once you have a large candidate pool, how do you select the best employee? Certainly there are basic standards they must meet. Do they have essential skills in the area? Do they have the necessary education and experience? Once you do your first cut, you are left with a group of people who look fairly equal—on paper.

Now it is the time to select the best one from this group. One of the best ways I have found to select excellent employees or students is to call their references. A letter of reference can hide the true feelings of the writer. By telephoning the references and asking them polite but pointed questions about the candidate, one can often get a better evaluation of the applicant. One of the most revealing questions is "Would you hire this person again?" But do not stop with the references the applicant provided. If there are those who you respect and trust working in the agency where the person is now employed, you can ask them for a candid evaluation of the person as well. You can ask about the candidate's interest level, knowledge, and ability to carry things through to completion. However, there is one skill that you should be concerned with above all others.

How well does the person get along with people? A group of eighty Berkeley Ph.D.s in science underwent a battery of personality tests, IQ tests, and interviews in the 1950s, and were then reevaluated in the 1970s. The study found that social and emotional abilities were four times more important to their success than their IQ.[5] The angry, sullen worker is like a stone thrown into a pond. Their ripples affect all other workers, reducing morale and productivity. Avoid these individuals at all costs! How can you judge whether a person will work well with other employees? Here your references can help.

Former supervisors may be reluctant to comment on the personality of an employee, but it can be easy to tell how well the worker got along with others if you know what to look for. An outstanding employee will usually be given a reference like "Wow, John is outstanding. He was one of the best employees I ever had, and I would hire him back in a second." Conversely, a mediocre employee receives a reference like "Hmmm, John is okay. I mean he was a safe-enough worker. Sometimes he ran into some personality issues with his coworkers, but he seemed to work out most of the time." Almost never will you get a reference that says "John was a bad employee. He worked horribly with others." Therefore, it is important to pay attention to subtle clues and voice inflections when asking someone about a former employee or coworker. You should also pay attention to regulations in your area regarding what you can or cannot ask others about the applicant.

Is it better to hire someone with the necessary job skills but is unmotivated or works poorly with others, or a person who is motivated, smart, with outstanding personal skills, but not currently qualified for the position? Clearly it is best to be skilled in both areas. However, if you can choose only one, Nordstrom, the successful retailer from the Pacific Northwest, hires motivated, personable applicants. They say they can train employees in the skills they need for the job, but it is hard to train them to get along with others.[6]

HOW TO MANAGE STAFF

Once you hire a top person, your next step is to keep them functioning at the highest level possible. If you discuss clear goals and expectations of the person at the beginning of employment, the employee knows what is important to you, and can work toward it. Lee Iacocca, successful former CEO of Chrysler Corporation, used a three-month work plan with his employees.[7] At the beginning of a three-month period the employee would write what he or she would like to accomplish in the next three months. Iacocca would read the plan and then negotiate what else should be included. Then, three months later, Iacocca and the employee would review what had been accomplished. If things did not get done, it was hard for the employee to make excuses, since he or she wrote the plan!

If you think you know what motivates employees, results of research on the subject might startle you. A study of 1,500 medical technologists by G. H. Graham and J. Unruh[8] ranked the following as the top five motivators:

Manager personally congratulates employees who do a good job.
Manager writes personal notes for good performance.
The organization uses performance as the major basis for promotion.
The manager publicly praises employee for good performance.
The manager holds morale-building meetings to celebrate successes.

Also, the majority of the most-effective motivating techniques were initiated by individual managers, not the company as a whole.

According to Graham and Unruh's research, it is apparent that appreciating an employee, both publicly and privately, is huge. Graham and Unruh concluded that appreciation practices based on performance, such as praise, are much more effective than appreciation practices based simply on presence of employees, such as turkeys or bonuses for everyone on the holidays. But Graham and Unruh were not the only researchers who found that appreciation was important.

William Peck investigated factors that motivated technical employees and leaders. He found both groups wanted to be regularly recognized for good performance, a feeling of belonging, to be involved in decision making, and opportunities for professional development.[9] Kenneth Kovach discussed two surveys of what employees wanted from their work, one conducted in 1946, and one conducted around 1979–1980. Although they were conducted thirty-five years apart, the results were similar. *Supervisors* ranked good wages as the highest motivator for employees in both studies. However, *employees* ranked full appreciation of the work done, and a feeling of being in on things, higher than job security and good wages in both.[10]

Appreciation and recognition of employees by the manager are inexpensive and effective motivators, and seem to be universally important. A competitive salary is required for staff satisfaction; however, many report that money motivates only in the short term. Over time, feelings of entitlement take over, and money quickly loses its ability to motivate.[11]

Being an advocate for your employee is another excellent way to build morale. Supervisors and their staff at the U.S. Geological Survey Cooperative Units Research Program consistently provide employee training opportunities, and work hard to provide funding for individual unit programs, even when budgets are tight. They took budget cuts within their own programs before passing these cuts down to their employees when the federal budget was reduced. Because of their strong support of their employees, these managers have been regarded with respect by unit personnel.

Management of the work process, not the end result, can negatively affect employee morale. Sometimes management of the process is important. DNA samples must be run in a particular way or they are invalid. Safety processes must be followed to the letter. However, for many tasks, giving employees an outline of the final product you seek, and coaching them during the process if they need help is an effective way of keeping them happy and getting things done. The famous quote by General George S. Patton says it all: "Never tell people how to do things. Tell them what to do and they will surprise you with their ingenuity!"

Process management rears its ugly head in many agencies in the form of constant reorganization.[12] Agency heads, or commissioners, have finite terms and the top management of natural resources agencies can rotate every four to six years, depending on changes in elected officials. When a new director takes over an agency, he or she can face incredible pressure to "do something." Often directors will undertake a reorganization to show how active they are to their commissions or legislature, but unless there is extreme dysfunction in the organization, the reorganization may cause more harm than good. An agency with which I was familiar was subject to regular reorganization. Management would be centrally located in head-quarters for a couple of years, and then supervisory authority would be moved to the regional offices. A biologist gave me the best explanation of what happens to the agency under constant reorganization. He said, "Scott, no matter how the agency is organized, I know who to go to to solve par-ticular problems, and how to get things done. When they reorganize, I am totally unfamiliar with whom to go to now, and everyone must relearn new jobs and responsibilities." When an agency is concentrating on reorganiza-tion, they are not concentrating on resource management, and all the time and energy that could be devoted to serving constituents is now channeled into figuring out who does what and defining territories. Although reor-ganization can look like an easy and effective way to give the appearance of "doing something," management might consider whether they are doing more harm than good.

How do you motivate an employee who is not completing tasks? You can praise the employee both publicly and privately if he is doing things right, but when you need to talk with an employee about improving some aspect of his performance, have that conversation privately so you do not embar-rass him. Additionally, how you ask your employee to do something can affect her enthusiasm for carrying out the assignment. An employee is more motivated to complete a task if she is asked to "head it up" or "take charge" of it, rather than just being told to do it.

Time management expert Alan Lakein suggests that if an employee is not completing a task, ask if he would like you to provide advice.[13] Then provide advice, and break up tasks into small segments to monitor progress. For example, say you asked your employee to finish a draft policy for trout management in the agency. It has been dragging for weeks, and it seems as if it will never get it done. Here is how you might approach the problem:

You: "I would break up the trout policy into these segments: high mountain lakes, lowland lakes, streams, rivers, and future research. How long would you need to do one section?"

Employee: "I think I could finish a section in a week."

You: "Okay, a week. What about this—can you start on streams on Monday? On Monday could you write up stream habitat restoration, Tuesday, stream angling regulations, and Wednesday, stream stocking procedures? I will check back with you on Thursday, and I will edit what you wrote. Plan on incorporating my edits Thursday afternoon, and you will be done on Friday and ready for the next section Monday morning."

Employee: "Okay, I think that would work."

Here the employee, not you, has suggested the amount of time he needs, so because of the commitment and consistency principle he should be motivated to finish in the time allotted.

A task might not be completed because the employee is unclear on her level of authority. Explaining the task and the level of authority she has in completing the task can help her carry through. There are several levels of authority that an employee can have. For example, say your department has to develop timber harvest schedules. You can ask her to do one of the following, each with an increasing level of authority:

• Research the topic and find options. Bring me back a list of options and I will make the decision on how to proceed.

- Research the topic, find options, and make a recommendation on how you think we should proceed. Please bring me this recommendation, and based on this and other information, I will make a decision.
- Research the topic, find options, and you (the employee) decide how we should proceed.

Take the Advice of Successful Supervisors! Seek Out Mentors!

I always seek out successful people for advice—especially those who are talented supervisors. Dr. William Shaw was the highly respected program leader of Wildlife and Fisheries at the University of Arizona. He was responsible for managing a staff of professors, which is no small feat—professors are notorious for their egos, and cannot be fired unless they do something very serious. When I was to become a leader of a unit attached to this program, which included helping to hire and supervise a couple of professors, Bill Shaw stopped by my office. He said, "Scott, I do two things when hiring professors: hire the best you possibly can, and then get out of their way!" I have used this simple, yet valuable, advice to successfully manage our unit's professors and high-level staff who are highly independent and incredibly skilled.

Dr. Wayne Hubert has regularly won widespread acclaim as a fisheries scientist and U.S. Geological Survey Cooperative Fish and Wildlife Unit Leader. When I first started work, he asked me to visit him for a few days in Wyoming. He gave me invaluable advice on how he ran his program, including some especially good tips about graduate student management. He asked graduate students to write a project plan at the beginning of their study. Once the plan

(continued)

Take the Advice of Successful Supervisors! (continued)

was completed, it was sent around for signatures from members of the graduate school committee supervising the project. This plan gave the student a direction toward which to work, and ensured all committee members agreed with the graduate student's project tasks to avoid surprises or conflicts later.

Dr. Paul Krausman, an expert on large mammals and once on staff with NASA's unmanned space program, produced a huge amount of quality research, including numerous articles and textbooks. He gave me advice on supervising a team working toward a project goal. He did not wait until the end of the project to see if everyone completed their work successfully. If you got to the end, and the work of one person was not sufficient, it could hold up the entire project. Therefore, he asked project contributors to provide examples of their work throughout the project. For example on edited books, chapter authors provided outlines first, then chapter drafts. If a person could not finish an outline by the due date, or was producing substandard work, he would be replaced before too much damage was done. This allowed him to successfully complete on time complex projects involving many contributors.

Supervisors who are successful and experienced can provide you with valuable tips to help you run your program and supervise your staff.

GETTING RID OF THE PROBLEM EMPLOYEE

Terminating an employee who is unsuited for his or her job is one of the most difficult but necessary actions a caring manager must make. If you spend a lot of time hiring good employees, hopefully, the time you spend firing bad employees will be minimized.

If you decide to terminate someone, it needs to be done quickly in order to minimize the effect on the rest of the workers. Poor performance from one worker can affect morale, and can spread to other employees if they believe you are unwilling to enforce workplace standards.

The first part of being able to remove a problem employee is to know the procedures of your organization in regard to terminations. How much notice does a person receive? What are the procedures of the actual termination? Who do you need to notify?

When problems start to arise with the employee, notify your supervisor immediately. When I was a young supervisor, I thought I should handle problems on my own, and the boss should not have to be involved. Things escalated with a problem employee, and my boss found out about the problems from others. He was not happy about this course of events. Later, when I notified my boss of all problems and kept him informed throughout the entire process, I found I had a strong, supportive advocate on my side.

It is usually not a surprise to employees when you tell them that they are doing poorly. Possibly they are not a good fit for their particular group of responsibilities. According to Lee Iacocca, if they are writing a three-month plan, and if they regularly do not complete assignments, they often have the sense that they are not being successful in their position.[14] I supervised a graduate student who I liked personally a great deal, but was having major problems completing deadlines and actively managing his project. I tried everything I could to coach him, but nothing seemed to work. After he took a vacation during an important part of his field season, I called him into my office and explained the problem. I told him I would like his resignation unless he could immediately improve his performance dramatically. He realized he was doing poorly and opted for resignation. I was honest when I told him that I thought the world of him, but we had to have results on the project. The termination proceeded as well as it could under the circumstances. I saw him in the hallway the day following his resignation, and to my surprise, he seemed quite happy. I think he knew he was in over his head and was happy to be out of the situation.

What if you get an employee who is belligerent and absolutely opposed to the termination? By carefully documenting all the things that the employee is doing wrong, you can provide evidence if the employee disputes your claims. Assembling others who have seen your employee's poor performance can provide support to your points if the employee appeals the decision. However, your most important responsibility is to double-check that you are clear about the regulations of your organization in regard to terminations so that you can avoid making a mistake that the problem employee can use to prevent his or her removal. Plan to get data, keys, and other sensitive material either before or at the time of termination, so that you can avoid losing important project data and potentially having work materials vandalized or stolen. Ensuring that your supervisor and your human resources staff is informed throughout the entire process will allow them to serve as your advocates, and prevent them from being surprised at a firing announcement.

"Management by Walking Around"

Peters and Austin, in their book *In Search of Excellence*, state that one of the greatest problems of American managers is that they are out of touch with their employees and customers. They often stay in their office without getting out and seeing what actually is going on. Peters and Austin recommend that managers regularly get out in the field with their employees; that they walk around in the laboratory to see what they are doing; and that they meet with customers on a regular basis. This "Management by Walking Around" allows the manager to hear customer and staff issues, complaints and compliments, and forges a stronger working relationship among all parties.[15]

WORKING WITH *YOUR* BOSS

Until now, we have discussed only how to manage employees. However, many more employees have supervisors than supervise others! How does one get along with supervisors? There are all types of supervisors with varying personalities. Many books have been written describing the various personalities of supervisors and how to work with them. I have been lucky to have superior supervisors much of my career. These supervisors have passed on important points to me. The following is a compilation of common ways to get along with your boss.

Be Good at What You Do

It sounds basic, but it is surprising how many people will try all types of different tricks to please their boss without just relying on basic competence. Being competent in what you do, and making sure your boss knows you are competent, is critical for happy employment. Pay particular attention to making sure your boss knows of your accomplishments, because if he does not know about it, he cannot reward you.

Treat Your Boss, *and* Your Coworkers and Employees You Supervise, with Respect

Some employees feel that if they just treat the boss well, that is enough. However, most bosses will want to know that the employee is well respected throughout the workplace, by both her coworkers and her employees. Word travels fast, and if you are unpopular with your coworkers or your employees, your boss will probably hear about it right away. If this happens, she will have to take time out of her busy day to investigate the allegations and deal with them. You do not want this to happen too many times if you can avoid it.

Be Upbeat

Nothing can bring down work-unit morale as quickly as someone who is constantly complaining about things. Someone with an upbeat, can-do attitude is often not only a superior employee, but influences others to have a can-do attitude as well. These are the types of employees that bosses will promote more rapidly and provide with more responsibilities.

Don't Surprise Your Boss

If a boss is blindsided with a problem from left field and embarrassed, he or she will hold you accountable, either consciously or unconsciously. A biologist in an agency I worked with ran afoul of a constituent. He did an excellent job informing me immediately, and I in turn informed my supervisor. Then when complaint calls started to come into the supervisor and me, she was ready and expecting it.

Work with Your Boss on Priorities

Sometimes employees will have a boss that gives them impossible deadlines or loads them up with way too much work to be completed in the allotted time period. How can one deal with this situation? Grumping in private about the excessive workload is not a productive solution. What works better is to ask the boss about her priorities. For example, if she hands you a new project while you are still working on the Jones Wildlife Management Area Report, say to her "I won't be able to get the Jones Area Report done if I start on this new project. I assume you would still like me to finish the Jones Report first, right? Or does this new work take higher priority?" If the boss is out of town, or otherwise cannot be contacted, some employees send them a note or email stating "I assume you still would like me to finish the Jones Report, and unfortunately I simply won't have time to work on both that and the new project now. Therefore, unless I hear from you that you would like me to prioritize the new project; I will finish the Jones Report first, and then get to the new project as soon as possible."

Make Your Boss Look Good. Be Loyal

Complaining about the ineptitude of a boss, no matter what you think, is a course of action that will rarely benefit you. It is also surprising how rapidly kind or vicious comments you make about someone gets back to them. Martin Van Buren, the "Red Fox of Kinderhook," so named because of the way he used his political skills, was the U.S. president who followed Andrew Jackson.[16] Andrew Jackson was popular and his support was important to whoever ran for president next. It is said that Van Buren told Peggy Eaton, a Jackson friend, that Jackson was the greatest man who had ever lived, with the strict understanding that the comment should not get back to Jackson. Of course Jackson found out about the comment rapidly, thought highly of Van Buren, and supported him for the presidency. It is equally common for a bad comment to get back to the boss. Well-respected employees only criticize their bosses under the most extreme circumstances. Striving to ensure the boss is presented in the most favorable light will build morale in the company, increase your standing, and make your boss happy.

So what if your boss is truly a jerk, someone you really cannot deal with at all? You stand more of a chance to change the boss's behavior if he or she respects you enough to talk with you on the side about the problem, perhaps over a cup of coffee, a beer, or lunch. I have been able to get along with almost every boss with whom I have worked (at least I have thought so!). However, sometimes it is impossible to get along with a boss, no matter what you do. In these cases, see chapter 10 on defending yourself against dirty tricks. You will either have to change positions, report the boss, or use some other tactic.

CONCLUSION

I cannot put it clearly enough—the people you work with can make or break your reputation. Take the time to hire good people, and once you have them on staff, do your best to make sure they know you appreciate them. If you hire good people and manage them well, you will receive many accolades and your work will be a major source of joy instead of one of constant frustration.

CHAPTER SUMMARY

- On research projects, staff time is usually the most expensive component, and in agencies, salaries are a major portion of the budget. Therefore prioritize hiring and motivating staff.
- Get the best staff by advertising for positions widely, working with your personnel office to see how to access the best candidates, and carefully checking credentials and references.
- Appreciation from the manager, either by written or verbal thanks, or public praise, can be one of the best ways to motivate employees. Being their advocate and telling them what you want done, instead of how they should do it, also improves their relations with you.

- To deal with problem employees, discuss problems in private, ask them if they would like you to provide advice, break up their tasks into smaller segments, and ensure that they know their level of authority to carry out tasks.
- If you need to terminate an employee, know the procedures of your organization regarding terminations, notify your supervisor and human resources staff, and then move swiftly. Most substandard employees know that they are having a problem and will resign if asked. However, to rid yourself of a belligerent employee, document poor performance and plan to get data, keys, and other sensitive material before termination.
- You will have a good relationship with your boss if you strive to be competent, upbeat, and to work well with your coworkers and employees. Do not surprise your boss, and do the best you can to make him or her look good.

CHAPTER NINE

How to Make a Good Impression in the Field

The guttural roar from the Colorado provided a steady background to the hot afternoon wind gusting through the cottonwoods and the high-pitched songs of a few birds flitting through the branches in their search for food. Shadows began to fall on the Kaibab Trail across the river, signaling the lateness of the day. The Kaibab was one of the two trails you could take down to this part of the canyon, and like the other, it could get terribly hot this time of year—usually over one hundred degrees. Accordingly, it was the site of some deaths, mostly due to dehydration, even though the chocolate brown rapids of the Colorado were nearby.

We had suspected Art[1] would have already arrived, and we were somewhat uneasy, even though Art was supposedly experienced. Art was from another agency, and was scheduled to work with us on a project at the bottom of the Grand Canyon.

In time, we saw a lone shape struggling down the trail, carrying a large bag. As it got closer, we realized it was Art. He arrived at the campsite carrying steaks, hard liquor, wine, but no water. He collapsed at camp, and proceeded to get drunk. We had loads of sampling gear to lug through the jagged canyons the next day—and in this country you wanted to be at your best. Unfortunately, Art spent the next two days either drunk or high, and was almost worthless. In fact, he was a danger and embarrassment to himself, the rest of the crew, and his employers. In this two-day trip, he had ruined both his and his agency's reputation with us, and alienated many of the other people he encountered at the bottom of the canyon.

There are few easier ways to destroy your reputation than to make a bad impression on field projects. Your fellow workers, bosses, and employees will all be watching to see how well you do in the field. Most stories about "bad" natural resources employees are based on field incidents. Many field skills are common sense and seem obvious; however, it is amazing how many "professionals" do not follow basic guidelines and risk their standing in the natural resources profession.

How does one make a good impression in the field? Using the communication skills presented elsewhere in this book can help things go smoothly. However, the unique demands of field work require additional skills, many beyond basic communication skills and natural resources topics typically covered in a university curriculum. The following list of field tips was gleaned from working with dozens of students and over one hundred employees, and describe actions (or non-actions) that can make or break your reputation. First we will discuss tips for everyone, and we will then relate those specific to team leaders or supervisors. Can the following be learned? You bet—I have violated most of them at one time or another, yet I follow most of them now. Hopefully, you will too. Many seem obvious!

TIPS FOR EVERYBODY

Perhaps you are a volunteer, helping with an established field project where you wish to make a good impression on a potential boss who is leading the crew. Maybe you are a lead supervisor, responsible for the numerous crews who survey soil condition over hundreds of miles of prairie. The tips presented below are important for all field workers, no matter what your level of responsibility.

Be Organized

The components of a large field sampling operation—whether you are a worker or a supervisor—are overwhelming. You must feed, house and transport workers and volunteers. You have to ensure all components of complex sampling gear, electrical gear, and communications gear are in place. You are required to have first aid and other safety equipment available. Being organized ahead of time with your equipment and sampling procedures is critical for smooth field operations. Natural resources professionals are often working in remote locations, where forgetting things is at best inconvenient and at worst hazardous. Being organized can save you considerable frustration and potentially avoid an accident.

A checklist to ensure that you have all needed materials before you go into the field is usually necessary. The pre-flight checklist has been used for decades by aircraft pilots responsible for ensuring the safety of their passengers and crew. It has no doubt saved many lives. When you develop your checklist, carefully go over the sampling procedure in your head and you can ask others on your team to edit your checklist as well. Then use the checklist when you are loading for the field, checking off each piece of equipment as it is placed into the truck.

Another trick to help you remember things, in addition to your checklist, is to go through the entire field procedure in your head before you leave. For example, I rarely forgot anything on scuba diving trips. Before I left, I would dress myself in my mind—first the farmer john wetsuit bottom, next the boots, then the hood, and the suit top. I would then attach the buoyancy compensator to the tank, and attach the regulator. I would strap my dive knife to my calf, and then lift my air assembly (b.c./tank/ regulator) onto my back. Next I would lift the weight belt and tighten it over my waist. The final steps were to place my mask over my head with the attached snorkel, and put my fins on my feet. For each step, I made sure each piece of gear was in the truck. If I did not remember putting the gear in the truck, I would go back and get it, and then start going over the procedure in my mind again where I left off.

Keep the Right Attitude

If you have the right attitude, you can turn even the most dreary, monotonous field work into a fun, rewarding activity. Use some of the tips below to make your field work as enjoyable as possible for you and others.

These Guys Needed a Checklist!

Natural resources professionals are not the only ones who need to use checklists and be organized in the field. January 8, 1815. Dawn broke on a ragtag American army unit huddled behind narrow earthen fortifications five miles downstream from the river city of New Orleans. On one side was the Mississippi River, the other a vast swamp. The only dry land passage to New Orleans was the dry spit of land occupied by American troops. The Americans consisted of a hodgepodge of Kentucky riflemen, pirates, and liberated Haitian slaves, under the direction of forty-seven-year-old Major General Andrew Jackson.

Marching across the open plane toward the Americans was a crack British army unit, including soldiers recently hardened in campaigns against Napoleon and those fresh from winning battles against other American forces further north. Under the command of thirty-six-year-old Major General Sir Edward Pakenham, the British forces were twice the size of the American forces.

All the British had to do was to smash through the American lines, take New Orleans, and block the shipping of the fledgling United States from leaving the Mississippi River, thus crippling the American economy. By all accounts the British should have taken the day. However, in the confusion of the British advance, they realized that they forgot the ladders needed to scale the American earthworks. The British could not advance because they had nothing with which to scale the fortifications, and were cut down below the ramparts by withering American rifle and cannon fire. The British suffered over two thousand casualties, while the Americans incurred only seventy-one. Because the British forgot the ladders and planned poorly in other ways, the Americans saved New Orleans from occupation. Andrew Jackson's reputation was made.[2]

Find Ways to Make the Field Work Fun

Carl Hubbs, zoology professor at the University of Michigan, was the first to seriously survey the Great Basin of Nevada for fishes.[3] For his 1934 expedition, he packed his Chevy sedan with canned goods, utensils, and bedding, and loaded collecting gear over the running boards. His family piled in, and they were on their way to survey the vast expanses of deserts for fish in tiny pools and small creeks. Along the way, on bone-jarring roads covered with dust and sharp volcanic rock, tires blew and were patched, five batteries were knocked loose onto the ground, and the punctured gas tank was sealed with chewing gum. As you might have imagined, Carl's three children should have rebelled under these conditions, but Carl and his wife Laura found ways to pacify them by making the field work fun. They gave them five cents for each fish species collected at different sites, and special awards for each new species (one dollar each) or genera (five dollars each) they found. This turned the hot, uncomfortable field work into an exciting competition for the children, just by putting a new spin on the situation.

If you are creative, you can put the same type of spin on your field work. Have friendly competitions to see who can survey the most sites (accurately, of course). Try to see who can tell the best story or who knows the most trivia. If the field work gets dull, activities like this can make the time fly.

Use Your Sense of Humor

A good, easygoing sense of humor is *so* important that I put it in a category all by itself. Our Inland Fisheries Research team at the Washington Department of Fish and Wildlife would conduct boat electrofishing surveys of northern lakes, and then stop the craft, float in mid-lake, and measure and pump out the stomach contents of the thousands of fish we caught. In the cold, light rain of a Pacific Northwest night, my hands would soon numb and tighten, losing all feeling. To process the catch, we would work most of the night as the wind whipped and stung our faces, plodding on like robots, wishing for a soft, warm bed. We had every reason to be miserable. Yet the people I worked with were easygoing, talented, and fun. They had a wonderful sense of humor, sharing funny stories and jokes without being condescending or petty. The attitude on these trips made me forget my discomfort, and I thoroughly enjoyed being there to complete these arduous tasks.

Do Not Complain

Peer pressure is not only something that teenagers experience. Peer pressure can also influence adults who will be watching the supervisors or coworkers set an example. Therefore, if someone complains, especially the supervisor, it will be highly contagious and spread rapidly through a work unit. Conversely, an upbeat attitude by supervisors or coworkers drives all to get things accomplished that are well beyond what is expected. Positive attitudes also attract the attention of the supervisors, who look for those who can be optimistic in the field for permanent hires or advancement.

Frances Hamerstrom was a famous naturalist and Aldo Leopold's only female graduate student.[4] Hamerstrom was raised in Boston to be a debutant and a socialite, but she would have none of it. She loved the outdoors and nature, much to the chagrin of her wealthy parents who thought that well-bred sophisticated society girls did not do that sort of thing. Every time she went to the dentist as a child, she was treated to a trip to Boston's Museum of Natural History. She loved the museum so much that she used a pencil point to jab her gums so she would have to visit the dentist, knowing that a museum trip would follow. She planted poison ivy along a path through the woods behind her house to keep her parents from finding out she had a secret garden filled with plants, grubs, worms, and other "fun" outdoor things. Even though she was used to the fine houses of the Boston area, she and her husband Frederick, also a naturalist, lived in leaking old shacks in a poverty-stricken area of Wisconsin for their field studies of the local wildlife. There was no indoor plumbing (she and Aldo Leopold—yes, that Aldo Leopold—dug a latrine out back), and her outdoor water pump froze in the winter; it once froze so hard that she had to wrap her fancy ballroom gown around it, douse it with kerosene, and light it on fire to melt the ice in the pump.

Frances could have complained: about her parents not wanting her to be a biologist, about the rickety old houses she and her husband occupied, about the days of long, arduous field work, about no indoor plumbing or running water after she grew up with the best Boston had to offer. However, she cheerfully ignored or found ways around these inconveniences, focusing on her larger goal, wildlife conservation. Before she passed away in 1998, she had studied under one of the most famous wildlife biologists ever, trained hundreds of young naturalists who remembered her with considerable fondness, and had successfully worked with others to save endangered birds from extinction. The inconveniences really did not seem to matter to a woman who led a full, wonderful life.

Respect Private Property

As a conservation professional, there is a very good chance that some of your field work will take place on private property. If you run afoul of a landowner, you probably will not be allowed on the property again, your boss will hear about it, or the landowner will form a bad impression of your agency.

You can protect the fragile landowner-agency relationship by first calling to say you are coming; by making sure closed gates are left closed and open gates are left open behind you; by staying on the road and not driving on fields or nearby land; by keeping noise to a minimum; by explaining to the landowner what you are doing; and by answering any questions they might have.

If you work long enough in the field you will probably do something that will irritate a landowner. You might get your truck stuck on his road, or accidentally break boards on his boat dock. If so, apologize profusely, and offer to repair any damage. We once got our trucks bogged down so deeply in the muck at a lake's edge that we had to work late into the night, digging and winching them out. The next day we surveyed the damage to the property, which was considerable. We called the landowner, apologized profusely, and then came back a couple of days later to fill in the ruts and smooth gravel back where needed. These actions preserved our friendship with him because it demonstrated we cared about his property and could be trusted.

Be First to Tackle the Toughest Tasks

In his book *Naturalist*,[5] E. O. Wilson discusses the character of famed zoologist Philip Jackson Darlington, curator in entomology at Harvard. Darlington had an outstanding reputation in the field, working long hours in the most difficult terrain imaginable to collect various insect species. His exploits were famous. He toiled and cut through the bush in some of the wildest places of Cuba, Colombia, and Haiti, scaling summits and squeezing through tree trunks, in some of the deep jungle that later hid guerilla bands fighting the government. In New Guinea and other Pacific islands he was present in World War II campaigns against the Japanese and retired a major, collecting insects during this time. His brush with death in a jungle pool was legendary. He fell off a log into the pool during his search for new species. A large crocodile grabbed him and took him to the bottom to drown him. An errant kick or a fortunate move allowed Darlington to escape the crocodile's jaws and swim toward land. As he climbed up the edge of the pool, the beast attacked again. Darlington fought and fought, screaming for help that did not come, and finally made it to shore. He then hiked through the jungle to a nearby hospital with the ligaments in both arms chewed and the bones in his right hand crushed. Did that stop Darlington from collecting insects? After a stay in the hospital, he headed out again, using his less damaged left arm to catch insects while his right arm healed.

As you might imagine, if Darlington asked you to do something, he had been there and done that. He was held in the highest esteem by his coworkers and students, all of whom knew that Darlington would not tell you to do something he would not be willing to do himself.

Never ask your workers to tackle something you yourself would not do. If you are part of the team, volunteer for the toughest jobs. Do something hard without being asked. To gain respect from others in the field, you should be willing to take on the hardest, most unpleasant tasks, and complete them with a smile.

Clean and Repair Borrowed Equipment

One of the most well-known stories in the history of the U.S. presidency deals with borrowing something and the consequences that followed.[6] The fact that this story is *so* well known should reveal to you how important taking care of borrowed equipment and other items can be for your reputation.

Abraham Lincoln was an avid reader throughout his life, but when he lived on the Indiana frontier, he found it exceedingly hard to get books. Books were quite valuable at that time, not common as they are today. If someone allowed you to borrow one, it was a big deal (imagine borrowing someone's laptop computer today). Lincoln borrowed a copy of Parson Weem's *Life of George Washington* from a nearby farmer and read it eagerly whenever he had free time. One night, he stored the book in between two cabin logs when he went to sleep, and a strong rainstorm badly soiled the book and warped its covers. He walked over to the farmer's house, explained the incident, and spent the next two days working for the farmer to pay his debt. How he handled this incident and others like it resulted in Abraham Lincoln getting the nickname "Honest Abe," a reputation that would serve him well in later life.

You too can develop an excellent reputation if you care for other's equipment even better than your own. When you return it, ensure that it is cleaner than it was when you got it and that you get it back on time.

Repairing borrowed equipment is also something that needs to be done in order to use it again and protect your reputation. If I broke a part in the field, such as hitting a propeller on a rock, I would replace it with money from my budget. Some things are not that clear cut. For example, say an engine breaks down in the field when you are using it, through no fault of your own. Sometimes staff can compromise on payment for repairs such as these. Perhaps each staff member pays half the cost of repairs out of their respective budgets. Usually the cost of repairs can be negotiated after a frank discussion. I borrowed another biologist's truck whose tires were almost bald. We shredded a tire when going out into the field. Some of the other biologists I worked with made the argument that the truck tires were already set to blow, so the full cost of replacement should be borne by the truck's owner. However, because I borrowed the truck, I offered to pay half and take it in and have it replaced. The cost of the tire was well worth the goodwill and respect earned by settling it this way. Repairing or sharing in the repair costs of equipment that you borrow will make it much easier to borrow the next time.

Move Quickly and Accurately When Working

A reporter described Dr. Ann "Mayfly" Morgan in the field as such: "a gray-haired lady knee-deep in muck and water ferociously pursuing nasty little herbivora and carnivora with a net. . . . she 'didn't ever want to come out' of some particularly oozy mudhole, where she happily captures nymphs, sponges and other creatures and encourages her students to do the same."[7] Somehow you cannot imagine Ann Haven Morgan, former professor and head of zoology at Mount Holyoke College in Massachusetts, ever letting moss grow on her.

Unfortunately Ann Haven Morgan's field behavior is not practiced by some. How often have you seen this happen? When some workers get to the field site, they might break out a sandwich and not get started right away. Work moves at a glacial pace. It seems to take forever to unload and set up the gear, and there are frequent breaks during the day.

If you learn how to make your body move quickly, and ignore minor discomforts, you will be surprised at how much you can accomplish. How else can you reduce the time needed to complete field activities without sacrificing accuracy? Identify where bottlenecks occur in field operations, and work to shorten the time needed to complete them. Pre-make data sheets before going into the field, have jars with ready labels, and ensure people know exactly what they will be doing before they reach the field site. Often the time required to complete a field operation can be halved or reduced by two thirds if rapid, accurate movement is stressed. You have the option of either rotating people on different tasks so they can try different things or allow a person to become proficient at one specific task that he does all the time. In my experience, allowing a person to manage the same task and master it during all field operations has been most efficient.

Prioritize Safety

Students, staff, and I were going to sample Burro Creek Canyon in the scorching deserts of western Arizona. I was a new professor in Arizona, anxious to explore the canyon. The first night, we camped in one of the canyon's tributaries, preparing for our several-hour walk into the main canyon the next day. Burro Creek Canyon contained only a few pools of water in midsummer, and throughout most of its length the canyon floor was dry. I packed three liters of water because the map showed a two-mile hike to the sampling site in the canyon. I decided not to bring the water filter because I did not think we would need it.

The heat of the Arizona summer is deadly to those who do not have enough liquids. Each year, hundreds die of thirst in southwestern deserts.

Because of the extreme heat, the water in all of our bottles disappeared quickly during the hike. We sampled the pool containing the fish, and I filled up my empty water bottles with stagnant creek water, just in case I might need it. We then hiked to a nearby road because we thought it would get us back to the car more quickly. However, because of the canyon's twists and turns, it took much longer to get to the road than we projected. By mid-afternoon we all ran out of water and the only thing we had to drink was the putrid, stagnant water in my bottles from the sampling station. We sucked on pebbles, tried to ignore our thirst on the way out, and finally got to the road. Next to the road was a small water puddle where cattle were standing and urinating. Our choice was to drink this or the stagnant water in my bottles. We were all set to drink one or the other when a car came around the corner. They gave us a ride, water, and asked, "Who is in charge of this operation?" I sheepishly raised my hand. Then they asked where we were from. I said, "The University of Arizona." They had a good laugh, because they were from Arizona State University, our arch sports rivals in the state! They said, "Wait until we tell people at ASU about you guys!" Although this situation was embarrassing, it ended well, with no injuries or anything worse. Since then I have always made sure I was well prepared for the field and have not had any more problems.

What factors are key for safety in field work? Make sure safety procedures are in place and personnel are well trained so there is less confusion if the unexpected happens. Desert safety classes are offered in Arizona. CPR and first aid classes are offered throughout the United States. Take classes like these to know what to do before something happens. I require my field staff to carry some form of communication, such as a cell phone. Communication allows you to get help rapidly in case of an emergency. Letting others know where you are going and when you will return helps them find you if you do not return at the appointed time. Sampling with others will give you backup help if you sprain your ankle or are dehabilitated in some other way.

Knowing the real hazards in your region allows you to focus and prepare for them. In Arizona, newcomers are commonly afraid of snakes, scorpions, and other venomous creatures, thinking they are the greatest danger. However, only about fifteen people per year die of snakebite in the entire United States.[8] As of 2000, the last fatality in Arizona by scorpion sting was in 1964.[9] Far more dangerous is the heat and dryness of the desert, which is responsible for thousands of deaths. Many of the same people who are deathly afraid of poisonous creatures think nothing of hiking in the desert with little or no water. In our program, we overkill on having enough water available. In the Washington State lake sampling program, the dangers were not lack of water or dangerous creatures, but hypothermia and armed, unstable individuals. We were prepared for both of these when we went sampling in Washington State.

Staying clear of alcohol and drugs goes without saying. It seems obvious, correct? However, the technician who was going to the bottom of the Grand Canyon to sample the creeks was experienced, yet drunk or high most of the time. Not only was he a danger to the others sampling, he reflected poorly on his agency.

TIPS FOR SUPERVISORS

A field leader needs to follow all the tips suggested above, plus those mentioned below for successful field work. The leader sets the tone in the field. A poor, complaining leader can affect the morale of the entire work unit, while a good leader can drive the team to accomplish things they never thought possible.

One of the best explanations of how a leader tackles tough tasks comes from the military. They say moving troops is like moving a piece of string. If the leader pushes from the back, the string just bunches up and goes nowhere. If the leader is in front and pulls the string, the string goes ahead readily in a straight line.

Arrive First, Leave Last

Arriving late when your employees and volunteers are waiting to begin is a poor way to start off your field work. Just getting to the office first is usually not enough to make a good impression. Having things packed before you meet and being ready to go can get your team into the field quickly and keep morale high. One way to ensure you will make a good impression is to load the field vehicles the night before, or if this is not possible, have containers of field equipment ready to load easily. Waiting around and leaving late on a field project can start the trip off with bad morale.

When I started work as a graduate student, I often showed up right on time or a bit late to go out into the field. Employees were rightly frustrated as we worked to load the truck and left late. As I worked to improve my skills, I loaded the truck and parked it in the locked compound the night before, ready to leave immediately the following day at the appointed time. I know I had "arrived" when a fellow student, a New York City native, known for his bluntness and honesty, told me he appreciated that we could leave right away, and not sit around and hang out while the biologist in charge worked on things he should have completed on the previous day.

At the end of the field day, making sure your field truck is cleaned out, gear is stowed, and all of your workers and volunteers have rides home will also enhance your reputation as a skilled field supervisor. At this time, most of the crew, including the supervisor, are incredibly tired. Your greatest temptation is to drive the truck into the lot and leave for home immediately so you can get some rest. At the Washington Department of Fish and Wildlife, experienced crew leaders fought this temptation. Upon arrival, they immediately started carrying field gear into the warehouse. Wet nets were hung to dry, boats were parked in the lot, and the truck was cleaned out. Crew leaders then asked volunteers and other workers if they had rides home, and if they did not, the leader gave them a ride. Once everyone else was taken care of, the crew leader left for home.

Demonstrate, Don't Tell

Robert Findley was a quiet supervisor, but one for whom I had tremendous respect. I was most fascinated by his leadership skills during my agricultural work in college. We were hoeing weeds in the hot, humid cornfields next to the Ohio River in southern Indiana. The corn was head high, yellow pollen was flying everywhere, the leaves were scratchy, and the fields trapped the stifling heat. Lunch under large shady trees at the river's edge provided a cool respite from the work. No one wanted to leave the trees after lunch to get back into the miserable cornfields. I was an occasional supervisor, and at the end of lunch I would say to the other workers "Okay, time to get back in the fields." Few would move when I first asked them, and they seemed to require more and more prodding each time. However, Bob Findley did not seem to have a problem. Instead of asking people to start work, at the end of lunch he would just get up without a word, grab a hoe, sharpen it, and start to walk toward the field. Others would see him, and then follow one by one back into the field. Not telling, but actually leading or demonstrating what to do, was the key to Bob's success as a field supervisor.

Once, at a meeting, I was trying to get biologists at an agency to snorkel at night for bull trout juveniles. Snorkeling had to be conducted at night because this was the only time the small fish would come out of the spaces in the gravel at the bottom of the river.[10] I overheard one of the older biologists remark, "BS, I'm not going to snorkel at night in those rivers. That guy's crazy." Another older biologist turned to him and remarked that he had seen me do it several times successfully in the rivers in his district. The fact that I had done it, knew what I was talking about, and had demonstrated that not only could it be done, but that it was the most effective way of sampling juvenile bull trout, stopped the criticism from the cantankerous biologist.

Treat Volunteers and Staff Like Gold

Field teams will often consist of two groups of people, volunteers and paid staff. Successful field professionals know to treat volunteers with the utmost respect. Biologist Bruce Bolding was so good at caring for volunteers that they usually wanted to sign up again. He recruited over a hundred volunteers for field projects, saving the Washington Department of Fish and Wildlife thousands of dollars in salary. Bruce would cook them dinner, have beers with them following field operations, and make sure that he had an extra set of field gear (raingear, boots, etc.) in case they forgot theirs. Cristina Velez and Laura Leslie recruited over 250 volunteers for their project, which examined fish populations of the Verde River, Arizona. They later wrote a publication on how to work with volunteers on fisheries projects.[11] They did many of the same things as Bruce, and volunteers were generally happy to come back. One of the best ways of keeping volunteers happy and getting them to come back is to appreciate them. Volunteers and employees from other departments and agencies do not have to work on your project. Sometimes they might be required by their bosses to help, but the quality of the job they do can be influenced by you.

Treat paid staff with considerable respect, just as you do your volunteers. Staff are the most expensive part of any field operation, and a good field technician is difficult to replace. Paid staff typically have more responsibility than volunteers, and of course one of their jobs should be to treat the volunteers with esteem. Other tips for working with your paid staff can be found in chapter 8.

CONCLUSION

By following these few simple guidelines, you can greatly improve your standing and reputation in your agency. A talented field worker is a must in any natural resources agency or private organization, and will be highly valued. Additionally, field work is usually quite fun, so the better you are at it, the more opportunities you will get to work in beautiful areas with energetic people!

CHAPTER SUMMARY

- It is easy to leave a good impression during field work if you follow a few simple steps.
- Arrive first at the field site and be the last to leave.
- Do not complain. Your attitude can bring the morale of the whole group down.
- If you demonstrate a task to people instead of just telling them to go do it, you can get people to do things much more easily.
- Using your sense of humor during field activities can make the day go much faster and can improve employee morale.
- If you are first to tackle the toughest tasks, you will develop the respect of your supervisors and team members.

- When borrowing equipment, make sure it is cleaned and in good repair when it is returned to maintain a good impression with your coworkers.
- To finish field work in a reasonable time period, move quickly and accurately when working.
- Being as organized as possible helps field operations go quickly and ensures accurate data acquisition.
- Your paramount task when you are in the field is to ensure your and your coworker's safety.
- Be respectful of private property. If you run afoul of a landowner, chances are you will not be allowed on the property again, your boss will hear about it, or the landowner will form a bad impression of your agency.
- Thank profusely those who help—treat your volunteers and staff like gold.

CHAPTER TEN

Defending Yourself from Dirty Tricks, Machiavellianism, and Other Annoyances

The note arrived in the midst of the fight, after we had scored a few points and were making some progress. We were trying to stop a developer from building on a mountaintop above the valley. The county regulations prevented him from building on most ridges; however, this ridge north of Tucson was not yet protected by these laws. This beautiful uplift in the Tortolita Mountains, covered with sun-varnished boulders, ocotillos, and stately saguaros, had stood undisturbed for millions of years. Standing on top of it during mid-summer, you could smell the fragrant creosote that told of impending rain, feel the moist wind, and watch the jagged lightning bolts cut across the southern sky against the backdrop of Tucson, the Santa Rita Mountains, and the Santa Cruz River curling down into Mexico.

As homeowners in the valley below, we were determined to fight the developer as best we could. An heir to a fortune, he did not need the money, and he certainly did not need to destroy this beautiful ridge to build winter homes for wealthy visitors from the north. We formed a group to halt the development, and soon leaders in the group emerged. One was another real estate developer, an extremely effective fighter. He knew all the laws and techniques to help stop the building on the ridge.

Our developer was so effective that the ridgetop development company fought back—but not fairly. One day, many of the neighbors received an anonymous note in their mail. It accused our developer of criminal charges in an unrelated land clearing case. The intent of the tactic was to smear his reputation and fracture the group from within. However, because of the clumsy way this technique was implemented, it had little effect on the group. Members even joked with the person, stating they could not believe they had to work with a criminal!

Most people negotiate and influence others using above-board methods similar to those we have discussed in previous chapters. These usually work well, and allow them to keep good relations with others. However, some individuals, in their quest for power, recognition, and status, use underhanded tricks to try to meet their goals. Occasionally these tactics might be tried on you. Therefore, it is important to learn how to defend yourself against them, even though these situations are rare. The good news is that the most effective techniques for fighting dirty tricks allow you to keep things positive and not become "dirty" yourself. In fact, some of the defense techniques are just an extension of the verbal judo methods we discussed in chapter 3.

What follows is a potpourri of methods you can use to identify and defend yourself from dirty tricks. Many were borrowed from the writings and biographies of the nation's top political consultants and politicians.[1] Others were gleaned from references on dirty politics, office politics, negotiation, and practical psychology listed at the end of this book.[2] These methods can be quite effective. Hopefully you will rarely have to use them.

HOW TO DEFEND YOURSELF AGAINST UNDERHANDED TACTICS

First, do not assume that someone is trying to use a dirty trick on you. Offhand comments, offensive remarks, or a bit of behind-the-back gossip or insults are to be expected in everyday life. Think of the times you made a thoughtless comment in passing and were sorry later. You are best to develop a thick skin and allow these to slide off your back. Experienced politicians are famous for this, and the most successful rarely hold grudges. However, if you repeatedly are attacked or if the attack is so strong to not be ignored, you may need to protect yourself.

For your first line of defense, use the techniques we have previously discussed. Have you tried to talk to the person, use verbal judo, or negotiate with him? If so, and nothing has seemed to work, move forward.

Before proceeding, it is important to realize why people use dirty tricks. People use dirty tricks to "get ahead" or destroy the competition. They do this by getting others to think poorly of you and well of them. They will not succeed if by using them, others think well of you and poorly of them. Because most people feel that the use of dirty tricks is underhanded, one of the best ways to fight dirty tricks is to expose them, and stay above the fray. This is done by explicitly confronting the person about the tactic, or by letting others know that the tactic is being used on you. You can use this simple idea to defend yourself against most dirty tricks. If you call attention to the tactic, the perpetrator will often be embarrassed, will know that he is discovered, and, in all likelihood, others will think poorly of him.

People who resort to dirty tricks are a bit like burglars. Burglars can break into most houses, but the trick is to make your house so impenetrable that it is not worth the effort. Similarly, you want to make it so hard for the person to perform dirty tricks on you that it is not worth the effort, and they will either give up or find someone else who is less skilled.

Character Assassination, Bullying, and Scapegoating

Sometimes a person tries to build themselves up by destroying your reputation. They might spread rumors, use you as a scapegoat, bully or blackmail you. What can you do about it? First, minimize your susceptibility to blackmail by being squeaky clean at work. Do not do anything that can be used against you. Give people who seem vindictive or have emotional problems a wide berth if possible and avoid confrontation with them unless absolutely necessary.

Despite our best efforts, most of us have been confronted with a character assassin, just like most of us as kids met up with a bully on the playground. By using some of the following techniques you can minimize their impact.

Exposing the Rumor

If someone spreads rumors about you, you can often stop the rumor by confronting him. An upbeat approach that attacks the problem head on can be effective. For example, "Mary, you seem to be a great person, intelligent, fun, and creative (see the verbal judo at work?). Therefore, I can't understand why people are saying you are spreading rumors about my work. Why am I hearing these? Are they true, and if so, did I do something to offend you? It's just completely out of character for you." This diffuses the situation by attacking the problem while still showing respect for the person (whether you feel it or not). Although they may deny that they did anything or they will have a weak excuse, the backstabbing will usually stop because the person realizes that you are on to her. She will pick on someone easier. If the behavior does not stop, then a more direct, forceful approach involving others may be necessary.

Documentation

If someone is consistently making life hard for you, document everything. A paper trail, detailed entries in a log, and even recordings saved from your answering machine (depending on policy or laws) will help support your points if you have to take the person to court, conduct disciplinary action at work, or save your reputation.

Networking

Building a network of friends among other coworkers, mentors, and managers will help buffer you against character assassination. For example, if Bill is bad-mouthing you, and Karen, Don, Tom, Jane, and Pete all know what a good person you are, Bill's word carries much less weight. I have been lucky to have had good supervisors throughout my career; however, friends of mine had to rely on their networks when attacked by a jealous, disagreeable boss. They had previously worked hard to build a network of powerful individuals such as other supervisors, top agency or university management, or leaders in their profession. When the supervisor, out of jealousy for the person's accomplishments, made their life hard or cut their position, they relied on this network of friends to pressure the boss to "ease up" or provide other job opportunities in extreme cases.

Inoculation

Do you feel that someone is about to bring up an issue to hurt your reputation? The effect of this can be minimized if *you* approach your supervisor, staff, or others about the issue first. For example, say you made a mistake on a permit application, making a constituent angry. Be sure to tell your boss before the constituent can. This reduces the surprise for the boss and often wins you points and an ally. A friend of mine worked for a state conservation agency in a position involving negotiation with industry representatives and legislators. One industry representative did not like the legislation that my friend was supporting and said to him, "I will ruin you politically." My friend could see what was coming. He went to the legislators as soon as possible and told them to expect to be contacted by the angry representative. The legislators were ready for the visit, and as expected, the industry representative came past their offices to attack my friend. One legislator said to the man, "I told (my friend) to work on this legislation. If you oppose him, you oppose me!" Needless to say, this was not the reception the industry representative was expecting. My politically savvy friend had stopped the character assassination by preemptively alerting the legislators of the impending personal attack.

Counterattack Hard—But Do Not Stoop to Their Level

Presidential political consultants are masters of countering attack ads and dirty tricks from the opposition. They recommend counterattacking hard, because just quietly taking the insults can make your critic appear right. A successful counterattack consists of three elements: (1) call attention to the attack, (2) defend yourself using objective criteria, and (3) stay above the fray. These methods can be best illustrated by an example. First is how *not* to deal with a negative attack. An ineffective counterattack, using insults against insults, can backfire.

Tom: Karen is an obstructionist bureaucrat. She really doesn't care about small business owners, and doesn't live in the real world. She and government employees like her really don't do anything worthwhile except take their government paycheck and tell people "no."

Karen: Well, I think Tom is a money-grubbing jerk, just out to tear up the wilderness so he can make a quick buck.

As you see, both have turned the argument into a "he said" "she said" confrontation. Someone witnessing this confrontation will probably not have sympathy for either side. Let us look at a more effective, positive technique for replying to Tom, one where Karen can fight back strongly, gain support from her coworkers and the public, and turn the tables on Tom.

Karen: I find it hard to believe that Tom has to resort to name calling (call attention to attack). He knows full well I have supported businesses who have worked hard to both make money and protect their property from environmental degradation, like Bigger's Grocery, Brandle Development, and Grinter Farms (providing objective criteria). Frankly, Tom is an effective business owner and I'm surprised he would resort to tactics like this. It is beneath him (staying above the fray).

Now Karen looks like an effective regulator who is above the fray, and Tom looks like a jerk. While not necessary, sometimes a compliment about your critic can boost your status as a fair, reasoned person. Karen's comment about Tom being an effective business owner would not be expected by her audience and would probably win her points.

The more you counter the attack in a dignified, objective, aggressive manner, yet staying above the fray, the better you look. Rachel Carson, author of *Silent Spring*, the groundbreaking book about the dangers of pesticides, counterattacked opposing scientists by publicizing objective information about the critic's funding sources—usually chemical companies or agricultural interests.[3]

Allow the Person to "Hang Himself"

John Douglas was a behavioral profiler of serial killers for the FBI and the model for Agent Jack Crawford in Thomas Harris's novel *The Silence of the Lambs*. Douglas wrote several books describing the mind of the serial killer and interrogation techniques. He was asked if by writing these books, he was giving away the techniques that were used to capture serial killers. Douglas replied that even if the serial killers knew how they were tracked, they still could not help themselves. Their aberrant behavior, which gives them away, is a part of their personality and not something they can easily stop.[5] Your nemeses are not serial killers (I hope), but this same principle works against them. People cannot help but show their "true colors" when exposed long enough to others.

If a person regularly exhibits poor behavior, he will not be able to turn if off in front of others and will ultimately "hang himself." Bringing others in to work with the individual as much as possible, or allowing them to see the individual in action, will often get you the support you need.

Exposing the Critic:
Fighting Senator Joseph McCarthy

Wisconsin Senator Joseph McCarthy achieved notoriety in the early 1950s, taking advantage of American's fear of communism by alleging communists had infiltrated many areas of American government and society. As chairman of the Senate Permanent Subcommittee on Investigations, he called dozens of hearings, interviewed hundreds of witnesses, and relentlessly grilled and insulted them. His smear campaign cost scores of careers, and he was greatly feared by many. On June 9, 1954, during hearings about communists in the U.S. army, lawyer Joseph Welch exposed Senator McCarthy's brand of dirty politics by explicitly calling attention to the tactics he was using at a public hearing. McCarthy attacked one of Welch's attorneys, claiming he had ties to a communist organization. Welch immediately counterattacked in front of a national TV audience. He told McCarthy, "Until this moment, senator, I think I never really gauged your cruelty or your recklessness." When McCarthy tried to continue his attack, Welch angrily interrupted, "Let us not assassinate this lad further, senator. You have done enough. Have you no sense of decency?" Overnight, McCarthy's career was destroyed. His national popularity disappeared, he was censured by the Senate and he was ignored by the press. He died three years later, a broken man, at age forty-eight. Exposing the "dirty politician"—yet staying above the fray—had worked.[4]

This was strikingly illustrated by Rachel Carson's interview on CBS in 1963 about her book *Silent Spring*.[6] Carson came under tremendous criticism from chemical and agriculture interests for her stance against DDT and other pesticides. She was painted as a hysterical woman, a communist, and an inexperienced reactionary by her critics. She appeared on the CBS interview as a well-dressed professional, reasoned and calm, objectively citing facts and figures about chemical pollution. In contrast, Dr. Robert White-Stevens, the chemical industry representative, appeared "wild-eyed and loud voiced," contributing exaggerated statements, such as "If man were to faithfully follow the teachings of Miss Carson, we would return to the Dark Ages, and the insects and diseases and vermin would once again inherit the earth." The poor showing of the chemical industry representative, who "hung himself," contrasted with the professional demeanor of Carson, further convincing the public of the validity of her claims.

Them's Fightin' Words!

One of the most colorful examples of character assassination occurred during the 1950 Democratic primary fight for a Florida senate seat between three-term incumbent Claude Pepper and challenger George Smathers. The campaign was described as vicious and extremely negative. Smathers depicted Pepper as a pro-communist apologist for Joseph Stalin, and a Leninist for supporting a national health-insurance bill.[7] In addition, Smathers supposedly took advantage of his constituents' lack of formal education by denouncing his opponent with statements that were true, but with big words designed to make Pepper sound wicked and immoral. Smathers declared Pepper was known all over Washington as "a shameless extrovert." His brother was a "practicing Homo sapien" and his sister "was once a thespian in wicked New York." He accused Pepper of practicing "nepotism" with his sister-in-law and "matriculating" with young women in college. He also stated that Pepper "habitually practiced celibacy" before marriage. These comments were reported in *Time* magazine and in northern newspapers, but Smathers denied he made the statements.[8]

Dealing with Those Who Steal Your Ideas and Take Credit

Theft of ideas can be problematic, especially in scientific fields. Leonardo da Vinci wrote his plans backwards in a mirror to prevent others from stealing them. Company laboratories are shrouded in secrecy to ensure that ideas are not stolen. A small business owner told me that secrecy is the most common way to protect ideas in business. If an idea is shared before the work is completed, it is frequently stolen.

Keeping ideas secret in natural resources is seldom possible or desirable. This is especially true if you collaborate with others, supervise staff, or provide regular updates to project sponsors. As a government employee, your work can be public information, by law available to those who would like to see it. Someone who deliberately keeps data from others can be looked upon as "uncooperative" and petty, especially when the data may help save an endangered species or habitat from destruction. But luckily, there are several ways to receive credit for your work when secrecy is not an option.

Finish First

Confederate Nathan Bedford Forrest was one of the most talented generals of the Civil War, reportedly driving Union General Ulysses S. Grant to fits of anger. His simple strategy was to get there first with the most,[9] meaning if you get to an area before the enemy with a larger number of troops, the battle is often yours. The same idea translates into protecting your ideas. Can you get there "first"? Move your idea to a final result as soon as you can. This can be finishing a publication, passing a regulation, or starting a new program.

Expose the Issue and Negotiate Over It

Politely and directly confronting the person who you think is stealing your idea as soon as you are suspicious can be effective. Do not automatically assume others are trying to steal your idea. They may not realize they are taking credit for your idea, or they will disagree with you about who should receive credit. If you first confront the person nicely about it, and you disagree who should get the credit, negotiate a solution. The issue will often be resolved. When I was a student, a collaborating professor wanted lead authorship on data I collected for my graduate degree. I felt strongly that because I collected the data, I should receive the most credit. I approached the professor with my concerns. He asked another professor for his opinion on the subject, and we negotiated a solution. I ended up getting lead authorship, but he presented the information at a national conference. This worked for both of us, and we preserved our good relationship.

Stake Out Your Territory

There is strong social pressure to not steal ideas in the natural resources profession. Those who are seen to plagiarize or "reinvent the wheel" do not get the type of reputation they seek. If your idea or project has already been exposed to limited circulation, especially among those who might steal it, you can do the opposite of keeping it secret. Quickly "stake out your territory" by getting a variety of sponsors to fund your project. It then becomes their project as well, and they can provide support if needed. Let a wide circle of friends, bosses, and associates know you are working on a particular idea. Take detailed notes of your work's progress. Email your bosses and peers regular "updates" describing your progress and include the person you feel might steal your idea in the mailing list. If the idea is widely known as yours already, it is less likely to be stolen by those who do not want to be seen as a plagiarizers or duplicators.

Focus on Collaborative Projects

Collaborate with others who are interested in the same subject area. Everybody wins in collaboration. You involve more minds to solve a problem so the solution is usually better. You spread the work among others. You involve those who might be tempted to take your idea if they are excluded. And you have a ready source of allies if someone does try to take your work.

Give Others Credit

Others around you will be less inclined to try to take credit for your ideas if you make sure *you* give them credit for their work and ideas. Frequently praising others and giving them credit can actually enhance your reputation. Making sure there is plenty of praise to go around sets an example for others, and gives people what they need—recognition.

Bureaucratic Intransigence

When our state agency research team wrote reports that were controversial, they usually ended up in exhaustive review, and were very difficult to publish. These reports went through numerous evaluations, and were never quite right to some of the middle managers and older agency biologists. In fairness, I cannot say these documents were held up on purpose. Controversial issues can be painful. People often procrastinate, hoping the issues will disappear, especially when there is a huge workload, such as for middle managers at a state agency. Some talented natural resources professionals are perfectionists, and will not finish a task until it is perfect. Others are so swamped with work that they cannot meet deadlines in a timely manner.

How can you speed up the glacial pace at which things often move in a bureaucracy? Here are some ideas:

Persistence
Many people want you to give up on controversial tasks because they upset the status quo. Be cheerful, considerate, and appreciative of their help, but continually ask if they have finished the task and are ready to move forward. Be the friendly pest and always keep things upbeat. The key is to push hard enough to get action, but not so hard that you infuriate the delayers. If you push hard with a big smile, you make it difficult for them to get angry at you.

Separate the People from the Problem, Which Is Delay
Compliment the delayer, but focus on the problem: "Mike, you are an excellent scientist and I really want to incorporate your comments. I'm sure you'll understand that I am under a strict timeline to get this done. Can you help me out?"

Let Them Suggest a Specific Date, and Then Hold Them to It
"When can you get this back to me?" If they suggest a date and promise to complete it by then, call them back the day it is due to check progress. Because of the commitment and consistency principle, they will be uncomfortable if they cannot produce it by the promised date.

Bypass Those Holding You Up
Say Mary has not yet edited your report, even though you have asked her several times, and she previously agreed to have the report completed by a specific date, which has long passed. Bypass Mary and work with Tom and Elaine to edit and finish the report. If your boss is holding you up, use this tactic with care.

How Floyd Dominy Built His Headquarters

An extreme case of bypassing was demonstrated by the colorful and flamboyant commissioner of the Bureau of Reclamation during the 1960s, Floyd Dominy. The Bureau of Reclamation's headquarters were housed in a series of decrepit converted hangars, warehouses, and barracks outside of Denver, Colorado. Dominy protested frequently that his organization needed a new building, but could not get support within the General Services Administration, the arm of the U.S. government responsible for constructing public buildings. Reclamation could not build buildings, but it could build dams. Therefore Dominy had his new building authorized as a dam, and got funding for it through Congress on that basis. For years, the Dominy Building, originally authorized as a dam and not close to any water, was the only high-rise building in Denver.[10]

Expose the Issue
If the above techniques do not work, you can expose the issue. A biologist whose report was held up wrote his boss an email asking when he could expect it. He did it in a friendly way, but carbon-copied upper-level management. Upper management then contacted the boss and the report was edited quickly. Informing an outside group with similar interests about the dilemma is effective, but a bit like playing with dynamite. Managers or biologists, who have had their programs tinkered with, held up, or cut, sometimes quietly notify outside groups with similar interests so they can take

action. These groups include private conservation organizations, other agencies, or angler or hunting clubs. The groups then put pressure on the agency's upper management to get the job completed. This type of exposure is powerful and usually gets things moving, but your job can be jeopardized if the "leak" is traced back to you. Therefore, if you decide to use this tactic, exercise caution and be aware of the consequences.

Naysayers

Naysayers are people who say an idea will "never work." If you give them too much credence, you will not carry through on most of your ideas, especially those that are new and novel. Unfortunately students or new staff members often take the advice of these people extremely seriously. In the past, they have come into my office, feeling downtrodden, and reported that "so and so" said our project would never work. Often "so and so" can be an older scientist or staff member who seems wedded to the status quo. Their seniority can often intimidate the young student or staff member. Also, inexperienced individuals often give the ideas of people who are critical more weight than those who are optimistic or neutral. You can defend yourself against naysayers using the following techniques.

Critically Evaluate Your Idea

Judge whether the person is offering constructive criticism that might improve your idea or is just trying to shoot it down because they are naysaying. If it is naysaying, realize that this behavior is common. Ignore it, and move on with your idea. If someone else is dealing with the naysayer and asks for your advice, expose the behavior. I tell students or staff members that, pardon the cliché, "If I had a nickel for everyone who told me that an idea would not work, I would be a billionaire. Do not worry about it and forge ahead."

Involve the Naysayer in Your Project

Ask the naysayer for criticism on your project and compliment them for it, or involve them in an aspect of your project. A naysayer is often looking for attention, and can be one of your strongest supporters if involved on your side. Lyndon Johnson described this in earthy terms when discussing FBI Director J. Edgar Hoover: "It's probably better to have him inside the tent pissing out, than outside pissing in."[11]

Detail Why the Project Will Work

If your project is well thought out and you have a detailed plan to complete it, you stand a better chance of convincing the naysayer. At worst you may not convince the naysayer, but you will likely convince others. I gave an overview of a book project for two sections of the American Fisheries Society. For one, I had a detailed outline, but for the other section, I did not. As you might imagine, it was much harder to convince the section who was not provided with the details why they should support the project. Any details of the project should describe the payoffs versus the inconveniences. Of course the payoffs should greatly outweigh the inconveniences.

Allow the Naysayer Time to Consider the Project Before a Decision

Ensure you have provided the naysayer with enough detailed information about potential payoffs. Following this, being patient and allowing the naysayer to mull over the project or idea for a while is often enough to get him to admit that it might work.

Bypass the Naysayer

If you cannot convince the naysayer, you may have to bypass him or her to get the project done another way. My graduate students had to build a recirculating aquarium system with computerized water temperature controls. A foreman employed at the facility where the system was to be built said "it would never work" and continually cast doubts on the students' abilities to finish the job. They ignored the naysaying comments of the foreman, bypassed him, and cheerfully worked with other staff members to finish their project. When the system was successfully completed, the foreman actually held them in begrudging respect.

In summary, listen to the naysayer and consider his argument. However, if you feel your idea has merit, forge cheerfully and politely ahead.

Lying

While most of us tend to believe people and take them at their word, sometimes people lie. It is extraordinarily hard to spot a liar. Most people correctly recognize a lie only 50 percent of the time. They might as well flip a coin. Federal law enforcement agents can increase the odds to 80 percent.[13] How do you interact with someone you believe to be lying? Fisher, Ury,

And You Think You Deal with Naysayers? Good Thing We Didn't Listen to These Comments!

"While theoretically and technically television may be feasible, commercially and financially I consider it an impossibility, a development of which we need waste little time dreaming."
—*Lee De Forest, U.S. inventor and "Father of the Radio," 1926.*

"That is the biggest fool thing we have ever done . . . The bomb will never go off, and I speak as an expert in explosives."
—*Adm. William Leahy, U.S. Navy officer, speaking to President Truman about the atomic bomb, 1945.*

"Everything that can be invented has been invented."
—*Director of U.S. Patent Office to President McKinley, 1899.*

"In the opinion of leading medical experts, milk given without medical supervision can do positive harm."
—*Alderman Arthur Barret, c. 1970.*

"Heavier-than-air flying machines are impossible."
—*Lord Kelvin, president, Royal Society, 1895.*

"Airplanes are interesting toys but of no military value."
—*Marechal Ferdinand Foch, professor of strategy, Ecole Superieure de Guerre, 1911.*

"There is not the slightest indication that nuclear energy will ever be obtainable. It would mean that the atom would have to be shattered at will."
—*Albert Einstein, physicist, 1932.*

"I see no point in reading."[12]
—*King Louis XIV.*

and Patton give several tips on how to deal with potential lying when negotiating.[14] These tips can be used for other situations as well.

Keep Negotiation Independent of Trust
If you sold your car to a stranger, you would not let someone drive it away on the promise that they would pay you later. The same idea holds true in negotiation. If you do not know someone or have a good reason to trust them, do not automatically assume they are telling the truth.

Verify Assertions of Fact
You can find out if a person is telling the truth by verifying their comments with others or by examining reports, letters, or memos related to the issue. For instance, supervisors who do not call references to confirm claims on resumes allow applicants to "fudge" their qualifications, which may result in substandard hires. By verifying claims on the applicant's resume with former supervisors and coworkers, the supervisor can detect lying.

Do Not Let Others Treat Your Need to Verify as a Personal Attack
Do not feel bad if you question the facts presented by someone. For example, if Janice is upset that you are questioning some information on her resume, you might say, "Janice, I know you wouldn't lie to me, but it is company policy that we verify the claims on all applicants' resumes."

Dirty Tricks During Negotiation

In *Getting to Yes*,[16] Fisher, Ury, and Patton describe several dirty tricks used in negotiation. The other party can refuse to negotiate; they can start out with extreme demands; they can use threats and personal attacks; or they might put you in a stressful situation. Fisher, Ury, and Patton relate that most people either try to ignore the dirty trick and move on with the negotiation, or they respond in kind. However, these responses are usually ineffective.

Fisher, Ury, and Patton recommend the best approach for dealing with such tactics is to call attention to them and negotiate over their use. For example, say the other party frequently interrupts you. You could ignore it, but the tactic may intensify and it might draw your mind away from the negotiation, causing you to settle for less. If you respond in kind, by also

How to Spot a Liar

Jack Trimarco is a former FBI interrogator, polygraph examiner, and profiler. He has conducted training on interview and interrogation techniques for the FBI Academy, the CIA, the U.S. Attorneys offices, the U.S. Department of Justice, INS, and numerous other state, federal, and local agencies. He currently hosts his own show on lie detection for Court TV. Below are ten common characteristics of liars according to Trimarco.[15]

- Liars deviate from their normal behavior. A liar who speaks with pauses will suddenly become a rapid talker.
- Liars are evasive. They will try to change the subject or gloss over a key point.
- People often pause when lying, giving them time to invent a story.
- Liars are not proud about being deceitful and unconsciously lower their voice.
- A liar's verbal and non-verbal behavior may be in conflict. They may say "no" while nodding "yes."
- Liars repeat questions when asked, trying to buy time to think of an appropriate response.
- Liars deny specific aspects of a crime. They won't admit to stealing $10,000 if he or she stole $9,999.
- Liars avoid eye contact—they don't want to see the target of the lie.
- A lie is hard to remember. Liars change their story over time.
- An innocent person's denial gets stronger over time, and they get angry. Liars don't get upset. They want to convince you that you are wrong about them.

interrupting the other person, you run the risk that negotiations could break off. The third option, calling attention to the trick and negotiating over its use, might work something like this:

"Tom, it seems as if you keep interrupting me while I speak. I don't think this is going allow either of us to negotiate a settlement that has the greatest chance of meeting both of our interests. How about if we set some ground rules that deal with our negotiation?"

WHAT IF *YOU* SCREW UP?

The above tactics are based on someone attacking you. But what if you screw up? A huge, monumental, colossal screwup. Both Richard Nixon and Bill Clinton screwed up royally during their presidencies. Nixon's staff members participated in the break-in of Democratic Headquarters at the Watergate Hotel, and Clinton had an affair with White House intern Monica Lewinsky. Political operatives Dick Morris, James Carville, and Paul Begala stated that Nixon would probably not have been forced to resign, and Clinton would not have been impeached, if they had admitted their errors immediately, apologized, and moved on.[17] It was not the act that resulted in their punishment, but the cover-up. People are immensely forgiving. Those who come clean early with a sincere apology may have to take some heat in the short term, but are quickly forgiven.

A University of Arizona graduate student made unauthorized changes to the school's website, harshly ridiculing another student for all to see. The graduate student was talented and showed promise, but had just screwed up colossally. He was rapidly discovered and confronted with the charges. Instead of trying to deny or excuse his actions, he immediately apologized to his supervisor, the ridiculed student, and other staff. Although he made a huge blunder, he saved himself by quickly apologizing and taking responsibility. Everyone was angry for a while, but the incident blew over, and the student is now doing fine. Therefore, if you screw up, admit your mistake as soon as possible and move on.

MACHIAVELLI AND *THE PRINCE*

When people talk about "dirty politics," the word "Machiavellian" often comes to mind. Some people think Machiavelli was an evil man and group him with Vlad III the Impaler, the Marquis de Sade, and Tamerlane. In truth, he was not particularly evil, but simply wrote about how politics were applied without considering morality. Most of his ideas were gleaned from studying the techniques of political figures who managed city-states during the Italian renaissance. These politicians often ruled with little regard for morality.

Machiavelli was an Italian diplomat who observed major politicians for about fifteen years. He belonged to a Florentine government that was deposed during a takeover by the rival Medici family. Because he was part of the previous regime, Machiavelli lost power. He tried to curry favor with the Medici family so they would let him back into government. He did this by writing down the successful political techniques he had observed during his fifteen years as Florentine secretary—a book of political advice for the Medicis. This was his masterwork, *The Prince*.[18]

Some of the advice *The Prince* offers politicians includes:

- It is better to be stingy than generous.
- It is better to be cruel than merciful (it is better to be feared than loved).
- It is better to break promises if it is against one's own interests.
- Princes must avoid making themselves hated and despised: the good-will of the people is a better defense than any fortress. (Let others do the unpleasant tasks.)
- Princes should undertake great projects to enhance their reputations.
- Princes should choose wise advisors and avoid flatterers.
- It is good to form strong alliances.
- Pamper people first. If you continue to have problems with them, crush them.

These principles were developed for ruling Italian city-states in the renaissance, so some have said that many of these principles are better suited to organizations ruled like Italian city-states than modern democracies. Those who rigidly adhere to all these principles usually include dictators, such as Hitler or Stalin, or Mafia bosses. For example, the practice of the Mafia bribing people to do something, and then killing them if bribery does not work, is an example of pampering people first and then crushing them. Even recent U.S. politicians apply some of these principles. In negative campaigning, groups who are "not connected" with a particular politician often spread vicious comments or air attack ads against their opponent. This allows the politician to act as if they never supported the attack, and stay above the fray. In natural resources, directors may have assistants that will do the "dirty work" of firing or disciplining people, while the director does not get overtly involved. How do you defend yourself against Machiavellian tactics? Become familiar with the tactics so you can recognize them as they occur. As with other dirty tricks, the next option is to call attention to the tactics, and negotiate over the effectiveness of their use.

Do you want to use Machiavellian techniques? Many current management books are written on how to apply dirty tricks or Machiavellianism. These techniques can work over the short term, but are rarely necessary if you are skilled at positive methods for presenting your points of view and rebutting your adversaries. If you develop a reputation for using Machiavellianism, it will hurt you in the long run.

CONCLUSION

Lee Atwater was a presidential political consultant who employed Machiavellian dirty tricks to win elections. At the end of his life, when he was facing an inoperable brain tumor, he had this to say about the methods he used:

My illness helped me to see that what was missing in society is what was missing in me: a little heart, a lot of brotherhood. The '80s were about acquiring—acquiring wealth, power, prestige. I know. I acquired more wealth, power, and prestige than most. But you can acquire all you want and still feel empty. What power wouldn't I trade for a little more time with my family? What price wouldn't I pay for an evening with friends? It took a deadly illness to put me eye to eye with that truth, but it is a truth that the country, caught up in its ruthless ambitions and moral decay, can learn on my dime. I don't know who will lead us through the '90s, but they must be made to speak to this spiritual vacuum at the heart of American society, this tumor of the soul.[19]

Machiavellianism can work, but often at a heavy moral price to the user. Keep your attack or defense fair, factual, and about issues rather than personal failings. You can be equally effective—and sleep well at night.

CHAPTER SUMMARY

- Most of us will encounter someone who uses dirty tricks at one time or another, just like most of us eventually ran into a bully on the playground when we were kids.
- Your first line of defense is to keep it positive. Have you tried to talk to the person, use verbal judo, or some of the other techniques discussed in this book? If these do not work, continue on to your defense against dirty politics.
- People most often use dirty tricks to get ahead, or destroy the competition. They do this by getting others to think poorly of you and well of them.
- A person will not succeed with their dirty trick if by using it, they get others to think poorly of them and well of you. Most people feel that the use of dirty tricks is underhanded.
- One of the best ways to fight dirty tricks is to explicitly raise the issue and expose those using them by confronting the person about the tactic, or by letting others know that the tactic is being used on you. Then stay above the fray. You can use this simple idea to defend yourself against most dirty tricks.
- Use the specific ideas presented in this chapter to protect yourself against character assassination, bullying, scapegoating, theft of your ideas, bureaucratic intransigence, naysayers, liars, and dirty tricks practiced during negotiation.
- If you are the one who screws up, admit it immediately and apologize.
- Defend yourself against Machiavellianism using the same defense for dirty tricks—by explicitly raising and confronting the person about the tactic; or by letting others know the tactic is being used on you. Using Machiavellian techniques can work in the short term, but often hurts your reputation over time. Use equally effective positive techniques.

CHAPTER ELEVEN

Conclusion

It really did not surprise anyone that it happened. It had happened before on many different occasions and in many other places. But the fact that people grew to expect it was the most disturbing of all.

On June 22, 1969, around noon, a train lumbered across the trestle over Cleveland's Cuyahoga River. The river was a putrid, fetid place, oozing and brown, covered with pollution and petrochemicals. The Federal Water Pollution Control Administration reported that the Lower Cuyahoga had no visible life, even low forms such as leeches or sludge worms. *Time* magazine ran a story that described the Cuyahoga using the following words: "Some river! Chocolate-brown, oily, bubbling with subsurface gases, it oozes rather than flows. 'Anyone who falls into the Cuyahoga does not drown,' Cleveland's citizens joke grimly. 'He decays.'"[1]

Some say the train had a broken bearing; others say it was something else. Whatever the case, a spark came off of the bridge and landed in the water. The spark ignited oil and kerosene, which in turn ignited floating debris.

The river "caught fire." The burn was not long or overly damaging and by the time the reporters arrived thirty minutes later, it was already out. It certainly did not cause the damage of other, earlier fires. The Cuyahoga had caught fire nine times before, the fire of 1952 resulting in $1.5 million in damage. The Cuyahoga was not unique in this respect. Other waters, such as a tributary to Baltimore Harbor, the Buffalo River in upstate New York, and the Rouge River in Michigan, also caught fire during this period.

In the 1950s and 1960s the state of our environment was not good. Poor water and air quality was common.[2] National media declared nearby Lake Erie "dead" because of all the toxic pollutants that had been dumped into it. Huge fish kills were reported in the Great Lakes. In New York, Los Angeles, and elsewhere, increased disease and death rates were linked to air pollution. In London, fog mixed with coal smoke killed approximately four thousand people over a four-day period in December 1952. In 1962, 750 Londoners died when a smoke-laden fog descended again. Mountains of litter lined our nation's highways. It was common to see people wing a bag full of trash out of their speeding cars. Many of the nation's fish and wildlife species were in peril. DDT had reduced the populations of many birds to extremely low levels, including our nation's symbol, the bald eagle.

Now, at least in developed countries, many things have improved. You would not expect a river like the Cuyahoga to catch fire, or see mountains of trash along our roads, or massive numbers of people dying in a pollutant-laden fog. Many species on the endangered list have either remained stable or have improved. As of this writing, bald eagle populations have rebounded. What happened? Why are things different now? Because in the 1960s and 1970s, individuals took action and made a difference. Some people worked alone, changing their own habits to favor environmental protection. They cleaned up their neighborhoods, recycled when possible, and stopped throwing bags of trash out car windows. Others made an even greater difference by working with other people and convincing them to adopt environmentally friendly habits, or by helping to pass laws. During the 1960s and 1970s, people used techniques similar to those demonstrated in this book to communicate the importance of natural resources protection to others.

Conservation professionals were instrumental in this tide of environmental cleanup and increased conservation. They helped to pass and enforce laws such as the Clean Water, the Clean Air, and the Endangered Species acts. They influenced and educated others about the benefits of environmental conservation. They led efforts to protect national parks, conserve wildlife, and restore streams, lakes, and oceans.

You do not have to examine history to see that strong government policies of environmental conservation and talented natural resources professionals can make an enormous difference. I have traveled around the world in countries where strong conservation programs were lacking. I saw factories belching tons of toxic waste into the air, litter so thick that you had trouble seeing the street, beaches and forest trails covered with human excrement, horrible over-population, and coastal fish populations that consisted of little more than a few, tiny individual fish. No one wants to live in an environment like that.

We still have a long way to go, all around the globe. The good news is that conservation professionals have excellent opportunities to continue to make huge changes to better our environment for future generations. We have every reason to be optimistic that we can do something. Unlike curing cancer, traveling to other universes, or fighting AIDS, we currently have much of the technology we need in place to protect the earth's resources and make our lives better. All we have to do is to decide to use it, and convince others to use it as well. Perhaps the greatest threat to our environment is overpopulation. Overpopulation can be controlled if people decide to adopt methods already in place to control reproductive rates. Pollution can be controlled if mankind decides to control its population and cut back on the production of pollutants. Species can be protected with the current technology if this is made a priority. Communicating with people and convincing them to adopt environmentally friendly practices is the greatest challenge facing the natural resources professional.

How can you make a difference? The skills presented in this book give you many of the basics for working more successfully with people. Other disciplines, such as business and marketing, are way ahead of the natural resources field in communicating their agendas. Learn from these disciplines. Books and articles can be found in the references section of book stores or your local library. Internet sites, seminars, and videotapes can also be valuable places to learn political skills. Also, learn from skilled communicators such as famous politicians, writers, and public speakers. Every day, new information becomes available on how to improve your skills in the social arena.

Few are experts at using these skills the first time they are applied, just as one would not expect to be an expert tennis player immediately after picking up a racquet. However, with persistence, these skills get easier. The best way to apply the skills in this book is by practice, and not berating yourself if you face setbacks. Persistent practice with these skills on a daily basis is key to improving your effectiveness.

Good luck to you in your pursuit of working more effectively with people! I am confident that your efforts to solve conservation problems will improve if you realize that working with people *is* a priority.

NOTES

Chapter 1

1. Not the real name of the lake. I still have friends that work up there!

2. Robert L. Wood, *Men, Mule, and Mountains: Lieutenant O'Neil's Olympic Expeditions*. (Seattle: Mountaineers Books, 1976).

3. Larry Heinemann, "Just Don't Fit—Stalking the Elusive 'Tripwire' Veteran," *Harper's Magazine*, April 1985.

4. *Washington Fact and Trivia* (2006 [cited December 17, 2006]); available from: http://www.50states.com/facts/washingt.htm.

5. William Dietrich, *The Final Forest: The Battle for the Last Great Trees of the Pacific Northwest* (New York: Penguin, 1993).

Chapter 2

1. Paul R. Ehrlich, "The Strategy of Conservation, 1980–2000," in *Conservation Biology: An Evolutionary-Ecological Perspective*, ed. Michael E. Soule and Bruce A. Wilcox (Sunderland, Massachusetts: Sinauer Associates, 1980).

2. Edward O. Wilson, *The Future of Life* (New York: Knopf, 2002).

3. Ibid.

4. United Nations Population Fund, "The State of World Population 2001. Footprints and Milestones: Population and Environmental Change," (2001).

5. Paul Ehrlich and J. P. Holdren, "Impact of Population Growth," *Science* 171 (1971).

6. George Pettinico, *The Public Opinion Paradox* (Sierra Club, 2005 [cited October 28, 2005]); available from http://www.sierraclub.org/sierra/199511/priorities.asp.

7. "Economists Challenge Bush Western Land Policies," *Reuters*, December 3, 2003.

8. Ernie Niemi, Ed Whitelaw, and Andrew Johnston, "The Sky Did Not Fall: The Pacific Northwest's Response to Logging Reductions," *Oregon's Future* (Summer/Fall 2000) 32–33; Associated Press, "Report: Spotted Owl Controversy Didn't Cause Overall Massive Job Losses," October 22, 1999.

9. Ernie Niemi, e-mail message to author, December 15, 2006.

10. *The Forum on Religion and Ecology* (2005 [cited December 16, 2005]); available from http://environment.harvard.edu/religion/main.html.

11. Associated Press, "Lichens Used in Research on Sierra Vista Leukemia," December 26, 2005.

12. Frank Stephenson, "A Tale of Taxol," *Florida State University Research in Review* 12, no. 3 (2002).

13. Homes Rolston, "Fishes in the Desert: Paradox and Responsibility," in *Battle against Extinction. Native Fish Management in the American West*, ed. W. L. Minckley and James E. Deacon (Tucson: University of Arizona Press, 1991).

14. Save Lake Davis Task Force Steering Committee and The California Department of Fish and Game, "Managing Northern Pike at Lake Davis: A Plan for Y2000," (2000).

15. Dennis P. Lee, "Northern Pike Control at Lake Davis, California," in *Rotenone in Fisheries: Are the Rewards Worth the Risks?*, ed. Richard L. Cailteux, et al. (St. Louis: American Fisheries Society, 1991); Ivan Paulson, *Pike Removal in Lake Davis, California* (2003 [cited December 16, 2005]); available from http://www.aquatic-invasive-species-conference.org/final_program_2003.htm.

16. Lee, "Northern Pike Control."

17. Jane Braxton Little, "What to Do About a Nasty Fish," *High Country News*, June 23, 1997; Save Lake Davis Task Force Steering Committee and The California Department of Fish and Game, "Managing Northern Pike at Lake Davis: A Plan for Y2000."

18. Paulson, *Pike Removal.*

19. Ibid.

20. Ibid.

21. Maria L. La Ganga, "Town Loses Fight to Stop Fish Kill-Off," *Los Angeles Times*, October 16, 1997.

22. Don Knapp, *Community Hopes to Kill California Plan to Poison Fish* (CNN Interactive, 1997 [cited January 16, 2001]); available from http://www.cnn.com/EARTH/9705/04/northern.pike/; Maria L. La Ganga, "Town Fights to Control Fate of Fish-Threatened Lake," *Los Angeles Times*, March 22, 1997.

23. Lee, "Northern Pike Control."

24. Dave Valle, *Davis Lake. Fall 97 Pike Poisoning* (1997 [cited January 16, 2001]); available from http://www.ecst.csuchico.edu/~jschlich/Flyfish/davis.lake.pike.shtml.

25. Associated Press, "State's Plan to Poison Lake Outrages Portola Residents," March 24, 1997.

26. Rodney Paige, "Hey Mr. Fish and Game." Song Protesting Rotenone Treatment of Lake Davis (1997).

27. Associated Press, *Fish Poisoning Plan Pits Locals against State* (1997 [cited January 16, 2001]); available from http://europe.cnn.com/EARTH/9710/14/poisoned.lake.ap/.

28. La Ganga, "Town Loses Fight."

29. Lee, "Northern Pike Control."

30. La Ganga, "Town Loses Fight."

31. Ibid; Associated Press, *Protesters Arrested as Officials Begin Poisoning Lake*(1997 [cited January 16, 2001]); available from http://cgi.cnn.com/EARTH/9710/15/poisoned.lake.ap/.

32. Maria L. La Ganga, "Lake Poisoning Settlement Gets Bogged Down," *Los Angeles Times*, August 28 1998, Maria L. La Ganga, "State Offers $9.1 Million for Lake Damage," *Los Angeles Times*, August 17 1998.

33. Sacramento Business Journal, "The Truth Floats Up" (May 8,1998 [cited January 16, 2001]); available from wysiwyg://49/http://www.bizjournals.com . . . ento/stories/1998/05/11/editoria12.html.

34. U.S. Fish and Wildlife Service, *Ash Meadows National Wildlife Refuge* (2005 [cited October 29, 2005]); available from http://www.fws.gov/desertcomplex/ashmeadows/index.htm.

35. Edwin P. Pister, "The Conservation of Desert Fishes," in *Fishes in North American Deserts*, ed. Robert J. Naiman and David L. Soltz (New York: Wiley, 1981).

36. James E. Deacon and Cynthia Deacon Williams, "Ash Meadows and the Legacy of the Devils Hole Pupfish," in *Battle against Extinction: Native Fish Management in the American West*, ed. W. L. Minckley and James E. Deacon (Tucson: University of Arizona Press, 1991).

37. Edwin P. Pister, "The Desert Fishes Council: Catalyst for Change," in *Battle against Extinction: Native Fish Management in the American West*, ed. W. L. Minckley and James E. Deacon (Tucson: University of Arizona Press, 1991).

38. Deacon and Williams, "Ash Meadows."

39. Ibid.

40. Pister, "The Desert Fishes Council."

41. Owen R. Williams et al., "Water Rights and Devil's Hole Pupfish at Death Valley National Monument," in *Science and Ecosystem Management in the National Parks*, ed. William L. Halvorson and Gary E. Davis (Tucson: University of Arizona Press, 1996).

42. Ibid.

Chapter 3

1. David D. Burns, *The Feeling Good Handbook* (New York: Plume, 1989); David D. Burns, *Feeling Good: The New Mood Therapy* (New York: Signet, 1980); Stephen R. Covey, *The Seven Habits of Highly Effective People: Powerful Lessons in Personal Change* (New York: Simon and Schuster, 1989); George J. Thompson and Jerry B. Jenkins, *Verbal Judo: The Gentle Art of Persuasion*, 2nd ed. (New York: Quill, 2004); Sam Horn, *Tongue Fu! How to Deflect, Disarm, and Defuse Any Verbal Conflict* (New York St. Martin's Griffin, 1996).

2. Burns, *Feeling Good Handbook;* Burns, *Feeling Good.*

3. Covey, *Seven Habits.*

4. Burns, *Feeling Good.*

5. Paul F. Boller Jr., *Presidential Anecdotes* (New York: Penguin, 1981).

6. Donald T. Phillips, *Lincoln on Leadership: Executive Strategies for Tough Times* (New York: Warner, 1992).

7. Burns, *Feeling Good.*

8. D. F. Gundersen and Robert Hopper, *Communication and Law Enforcement* (New York: Harper and Row, 1984).

9. PBS Frontline, *Ghosts of Rwanda* (2004 [cited October 29, 2005]); available from http://www.pbs.org/wgbh/pages/frontline/shows/ghosts/.

10. George J. Thompson and Jerry B. Jenkins, *Verbal Judo: The Gentle Art of Persuasion*, 2nd ed. (New York: Quill, 2004).

11. Burns, *Feeling Good*.

Chapter 4

1. Abraham H. Maslow, *Motivation and Personality*, 2nd ed. (New York: Harper and Row, 1970).

2. Marc Reisner, *Game Wars: The Undercover Pursuit of Wildlife Poachers* (New York: Penguin, 1991).

3. BBC News, *"Virgin Mary" Toast Fetches $28,000* (2004 [cited December 31, 2005]); available from http://news.bbc.co.uk/go/pr/fr/-/2/hi/americas/4034787.stm.

4. Robert B. Cialdini, *Influence: Science and Practice*, 4th ed. (Boston: Allyn and Bacon, 2001).

5. Theodore Roosevelt, "John Muir: An Appreciation," *Outlook* 109 (1915), 27.

6. Stephen Fox, *The American Conservation Movement: John Muir and His Legacy* (Madison: University of Wisconsin Press, 1985).

7. Cialdini, *Influence*.

8. American Friends Service Committee, *Uncommon Controversy: Fishing Rights of the Muckleshoot, Puyallup, and Nisqually Indians* (Seattle: University of Washington Press, 1970).

9. Robert A. Caro, *The Years of Lyndon Johnson: Master of the Senate* (New York: Knopf, 2002), xxii.

10. Cialdini, *Influence*.

11. Ibid.

12. Ibid.

13. David D. Burns, *Feeling Good: The New Mood Therapy* (New York: Signet, 1980).

14. Stephanie Barczewski, *Titanic: A Night Remembered* (New York: Hambledon and London, 2004).

15. Ibid.

16. James E. Deacon and Cynthia Deacon Williams, "Ash Meadows and the Legacy of the Devils Hole Pupfish," in *Battle against Extinction: Native Fish Management in the American West*, ed. W. L. Minckley and James E. Deacon (Tucson: University of Arizona, 1991).

17. Philip B. Kunhardt Jr., Philip B. Kunhardt III, and Peter W. Kunhardt, *Lincoln: An Illustrated Biography* (New York: Portland House, 1992); Carl Sandburg, *Abraham Lincoln: The Prairie Years*, 3 vols., vol. 1 (New York: Dell, 1954).

18. John Ed Pearce, "Harland Sanders: The Man Who Would Be Colonel.," in *The Human Tradition in the New South*, ed. James C. Klotter (New York: Rowman and Littlefield, 2005).

19. Stuart Ball, *Winston Churchill: The British Library: Historic Lives* (New York: New York University Press, 2003); Randolph S. Churchill, *Winston S. Churchill*, vol. 1 (Boston: Houghton Mifflin, 1966); Earl of Birkenhead, *Churchill: 1874–1922* (London: Harrap, 1989).

20. Sarah K. Bolton, *Lives of Poor Boys Who Became Famous* (New York: Thomas Y. Crowell, 1947).

21. American RadioWorks, *With This Ring: Following the International Diamond Trail* (2001 [cited October 28, 2005]); available from http://americanradioworks.public radio.org/features/diamonds/index.html; Costas Arkolakis, *De Beers: A Monopoly in the Diamond Industry* (2005 [cited October 28, 2005]); available from http://www.econ.umn.edu/~dmiller/diamondsPP.pdf; Tracie Rozhon, "Competition Is Forever: The Net and Big Sellers Change the Diamond Business," *New York Times*, February 9, 2005; Peter Verburg, "Diamond Cartels Are Forever. A Campaign to Stop Smuggled Diamonds from Funding African Warlords Isn't Purely Altruistic," *Canadian Business*, July 10, 2000.

22. Edana Eckart, *California Condor* (New York: Rosen Book Works, 2003); Jonathan London and James Chaffee, *Condor's Egg* (San Francisco: Chronicle Books, 1994); Patricia A. Fink Martin, *California Condors* (New York: Children's Press, 2002).

Chapter 5

1. Michael LeBoeuf, *How to Win Customers and Keep Them for Life* (New York: Berkley, 2000).

2. Zig Ziglar, *Zig Ziglar's Secrets of Closing the Sale* (New York: Berkley, 1984).

3. Alan Lakein, *How to Get Control of Your Time and Your Life* (New York: David McKay, 1973); LeBoeuf, *How to Win Customers and Keep Them for Life.*

4. Ken Blanchard and Sheldon Bowles, *Raving Fans: A Revolutionary Approach to Customer Service* (New York: William Morrow, 1993).

5. Tom Peters and Nancy Austin, *A Passion for Excellence: The Leadership Difference* (New York: Warner, 1985); Robert Spector and Patrick D. McCarthy, *The Nordstrom Way: The Inside Story of American's # 1 Customer Service Company* (New York: Wiley, 1995).

6. Thomas J. Peters and Robert H. Waterman Jr., *In Search of Excellence: Lessons from America's Best-Run Companies* (New York: Warner, 1982), 161.

7. Robert B. Cialdini, *Influence: Science and Practice*, 4th ed. (Boston: Allyn and Bacon, 2001).

Chapter 6

1. Mark Kurlansky, *Cod: A Biography of the Fish That Changed the World* (New York: Penguin, 1997).

2. Gail Bingham, *Resolving Environmental Disputes: A Decade of Experience* (Washington, D.C.: Conservation Foundation, 1986); J. Walton Blackburn and Willa Marie Bruce, *Mediating Environmental Conflicts: Theory and Practice* (Westport, Connecticut: Quorum, 1995).

3. Lawrence Susskind, Paul F. Levy, and Jennifer Thomas-Larmer, *Negotiating Environmental Agreements* (Washington, D.C.: Island Press, 2000).

4. Roger Fisher, William Ury, and Bruce Patton, *Getting to Yes: Negotiating Agreement without Giving In*, 2nd ed. (New York Penguin, 1991).

5. Ibid.

6. Information on Roosevelt from Joseph Bucklin Bishop, *Theodore Roosevelt and His Time: Shown in His Own Letters*, vol. 1 (New York: Scribner's, 1920); Paul F. Boller Jr., *Presidential Anecdotes* (New York: Penguin, 1981); Edmund Morris, *Theodore Rex* (New York: Random House, 2001). Roosevelt quote in: *Nobel Lectures, Peace 1901–1925 Volume 1* edited by Fredrick W. Haberman (New York: Elsevier, 1972), 100.

7. Susskind, Levy, and Thomas-Larmer, *Negotiating Environmental Agreements*.

8. Robert A. Caro, *The Years of Lyndon Johnson: Master of the Senate* (New York: Knopf, 2002).

9. Information on Vietnam negotiations from William Burr and Jeffrey Kimball, "Nixon's Nuclear Ploy," *Bulletin of the Atomic Scientists* 59, no. 1 (2003); Public Broadcasting System, *American Experience: People and Events: Paris Peace Talks* (2005 [cited April 1, 2005]); available from http://www.pbs.org/wgbh/amex/honor/people events/e_paris.html; Associated Press, "Nixon Had Notion to Use Nuclear Bomb in Vietnam" (*USA Today*, February 28, 2002 [cited October 29, 2005]); available from http://www.usatoday.com/news/washington/2002/02/28/nixon-tapes.htm.

10. Robert B. Cialdini, *Influence: Science and Practice*, 4th ed. (Boston: Allyn and Bacon, 2001).

11. Scott Mernitz, *Mediation of Environmental Disputes: A Source Book* (New York: Praeger, 1980).

12. Bay Area Historical Society, *The History of Silver Bay* (2004 [cited October 29, 2005]); available from http://www.silverbay.com/history.htm.

13. Stephanie Hemphill, *The Legacy of the Reserve Mining Case* (Minnesota Public Radio, 2003 [cited October 29, 2005]); available from http://news.minnesota.public radio.org/features/2003/09/29_hemphills_reservehistory/.

14. Mernitz, *Mediation of Environmental Disputes*.

Chapter 7

1. Alan Lakein, *How to Get Control of Your Time and Your Life* (New York: David McKay, 1973); R. Alec Mackenzie, *The Time Trap: How to Get More Done in Less Time* (New York: Amacom, 1972); Stephanie Winston, *The Organized Executive; New Ways to Manage Time, Paper, and People* (New York: Warner, 1983).

2. J. M. Juran, *Managerial Breakthrough: The Classic Book on Improving Management Performance* (New York: McGraw-Hill, 1995).

3. Lakein, *How to Get Control*.

4. Ibid.

5. The Aldo Leopold Foundation, *Aldo Leopold* (2005 [cited April 14, 2005]); available from http://www.aldoleopold.org/About/Leopold_bio.htm.

6. Philip Sterling, *Sea and Earth: The Life of Rachel Carson* (New York: Thomas Y. Crowell, 1970).

7. Winston, *Organized Executive*.

8. David D. Burns, *Feeling Good: The New Mood Therapy* (New York: Signet, 1980).

9. Ibid.

10. *Victor Hugo: Writer's Block* (2005 [cited October 28, 2005]); available from http://anecdotage.com/index.php?aid=18814.

11. Reuters, "Flight School Memo Named Bin Laden," May 15, 2002.

12. Winston, *Organized Executive*.

13. Burns, *Feeling Good*.

14. Ibid.; University of Michigan Faculty and Staff Assistance Program (FASAP), *Stress Manager* (University of Michigan Faculty and Staff Assistance Program, 2006 [cited January 14, 2006]); available from http://www.umich.edu/~fasap/stresstips/31.html.

15. Lee Iacocca and William Novak, *Iacocca: An Autobiography* (New York: Bantam, 1986).

16. Jane Goodall and Phillip Berman, *Reason for Hope: A Spiritual Journey* (New York: Warner, 2000).

17. John Gribbin, *The Scientists: A History of Science Told through the Lives of Its Greatest Inventors* (New York: Random House, 2004).

18. Shirley Streshinsky, *Audubon: Life and Art in the American Wilderness* (Athens: University of Georgia Press, 1998).

19. Curt D. Meine, *Aldo Leopold: His Life and Work* (Madison: University of Wisconsin Press, 1991).

20. Carol Grant Gould, *The Remarkable Life of William Beebe: Explorer and Naturalist* (Washington, D.C.: Shearwater Books, 2004).

21. Leonard Warren, *Constantine Samuel Rafinesque: A Voice in the American Wilderness* (Lexington: University of Kentucky Press, 2004).

22. Alex Ayres, *The Wit and Wisdom of Abraham Lincoln* (New York: Meridian, 1992).

23. National Institute of Mental Health, *What to Do When a Friend Is Depressed . . .* (National Institute of Mental Health, National Institutes of Health, U.S. Department of Health and Human Services, 2001 [cited January 14, 2006]); available from http://www.nimh.nih.gov/publicat/NIMHfriend.pdf.

24. National Institute of Mental Health, *Facts About Anxiety Disorders* (National Institute of Mental Health, The National Institutes of Health, U.S. Department of Health and Human Services, 1999 [cited January 14, 2006]); available from http://www.nimh.nih.gov/publicat/adfacts.cfm.

25. Stephen E. Ambrose, *Undaunted Courage: Meriwether Lewis, Thomas Jefferson, and the Opening of the American West* (New York: Simon and Schuster, 1997).

26. Pickover, *Strange Brains and Genius: The Secret Lives of Eccentric Scientists and Madmen* (New York: Plenum, 1998).

27. List of famous people with depression compiled from: Eli Lilly and Company, *Disease Information. Depression* (2005 [cited January 14, 2006]); available from http://www.prozac.com/disease_information/depression.jsp; The National Alliance on Mental Illness—New Hampshire, *Take Action—Famous People with Mental Illness* (2006 [cited January 14, 2006]); available from http://naminh.org/action-famous-people.php.

28. Healthyplace.com, *Famous People Who Have Experienced an Anxiety Disorder* (2006 [cited January 15, 2006]); available from http://www.healthyplace.com/communities/anxiety/paems/people/.

29. Burns, *Feeling Good Handbook*; Burns, *Feeling Good*.

30. Alan M. Langlieb and Jeffrey P. Kahn, "How Much Does Quality Mental Health Care Profit Employers?," *Journal of Occupational and Environmental Medicine* 47, no. 11 (2005).

Chapter 8

1. G. E. Hutchinson, "W. Thomas Edmondson," *Limnology and Oceanography* 33 (1988); J. T. Lehman, "Good Professor Edmondson," *Limnology and Oceanography* 33 (1988); Robert L. Wallace, John J. Gilbert, and Charles E. King, "In Memoriam: W. T. Edmondson (1916–2000)," *Hydrobiologia* 446–447, no. 1 (2001).

2. Hutchinson, "W. Thomas Edmondson"; Lehman, "Good Professor Edmondson"; Wallace, Gilbert, and King, "In Memoriam."

3. W. T. Edmondson, *The Uses of Ecology: Lake Washington and Beyond* (Seattle: University of Washington Press, 1991).

4. Jeffrey J. Fox, *How to Become a Great Boss: The Rules for Getting and Keeping the Best Employees* (New York: Hyperion, 2002).

5. Cary Cherniss, "Emotional Intelligence: What It Is and Why It Matters" (paper presented at the Annual Meeting of the Society for Industrial and Organizational Psychology, New Orleans, Louisiana, 2000).

6. Robert Spector and Patrick D. McCarthy, *The Nordstrom Way: The Inside Story of America's #1 Customer Service Company* (New York: Wiley, 1995).

7. Lee Iacocca and William Novak, *Iacocca: An Autobiography* (New York: Bantam, 1986).

8. Gerald H. Graham and Jeanne Unruh, "The Motivational Impact of Nonfinancial Employee Appreciation Practices on Medical Technologists," *The Heath Care Supervisor* 8, no. 3 (1990).

9. William F. Peck, "Motivating Technical Employees," *Civil Engineering* 64, no. 3 (1993).

10. Kenneth A. Kovach, "Why Motivational Theories Don't Work," *Advanced Management Journal* 45, no. 2 (1980).

11. Terry Bragg, "How to Reward and Inspire Your Team," *IIE Solutions* 32, no. 8 (2000).

12. Robert F. Carline, "The Folly of Reorganization," *Fisheries* 23, no. 10 (1998).

13. Alan Lakein, *How to Get Control of Your Time and Your Life* (New York: David McKay, 1973).

14. Iacocca and Novak, *Iacocca*.

15. Thomas J. Peters and Robert H. Waterman Jr., *In Search of Excellence: Lessons from America's Best-Run Companies* (New York: Warner, 1982).

16. Paul F. Boller Jr., *Presidential Anecdotes* (New York: Penguin, 1981).

Chapter 9

1. Not his real name.

2. Information on Battle of New Orleans from Marquis James, *The Life of Andrew Jackson* (New York: Bobbs-Merrill, 1938); Jean Lafitte National Historical Park and Preserve, *Chalmette Battlefield and National Cemetery* (Jean Lafitte National Historical Park and Preserve, 2005 [cited October 28, 2005]); available from http://www.nps.gov/jela/Chalmette-battlefield.htm.

3. Robert Rush Miller, Clark Hubbs, and Frances H. Miller, "Ichthyological Exploration of the American West: The Hubbs-Miller Era, 1915–1950," in *Battle against Extinction: Native Fish Management in the American West*, ed. W. L. Minckley and James E. Deacon (Tucson: University of Arizona Press, 1991).

4. Helen McGavran Corneli, *Mice in the Freezer, Owls on the Porch: The Lives of Naturalists Frederick and Frances Hamerstrom* (Madison: University of Wisconsin Press, 2006); Frances Hamerstrom, *My Double Life: Memoirs of a Naturalist* (Madison: University of Wisconsin Press, 1994).

5. Edward O. Wilson, *Naturalist* (New York: Warner, 1995).

6. Doris Kearns Goodwin, *Team of Rivals: The Political Genius of Abraham Lincoln* (New York: Simon and Schuster, 2005).

7. "Marcia M. Bonta, *Women in the Field: America's Pioneering Women Naturalists*, 1st ed. (College Station: Texas A&M University Press, 1991).

8. Michael P. Ghiglieri and Thomas M. Myers, *Over the Edge: Death in the Grand Canyon* (Flagstaff, Arizona: Puma Press, 2001).

9. Ibid.

10. Scott A. Bonar, Marc Divens, and Bruce Bolding, "Methods for Sampling the Distribution and Abundance of Bull Trout/Dolly Varden" (Olympia: Washington Department of Fish and Wildlife, 1997).

11. Laura L. Leslie, Cristina E. Velez, and Scott A. Bonar, "Utilizing Volunteers on Fisheries Projects: Benefits, Challenges, and Management Techniques," *Fisheries* 29, no. 10 (2004).

Chapter 10

1. Paul F. Boller Jr., *Presidential Anecdotes* (New York: Penguin, 1981); John Joseph Brady, *Bad Boy: The Life and Politics of Lee Atwater* (New York: Perseus, 1996); Robert A. Caro, *The Years of Lyndon Johnson: Master of the Senate* (New York: Knopf, 2002); James Carville and Paul Begala, *Buck Up, Suck Up . . . And Come Back When You Foul Up: 12 Winning Secrets from the War Room* (New York: Simon and Schuster, 2002); James Carville and Paul Begala, *Take It Back: Our Party, Our Country, Our Future* (New York: Simon and Schuster, 2006); Michael Dorman, *Dirty Politics: From 1776 to Watergate* (New York: Delacorte, 1979); Doris Kearns Goodwin, *Team of Rivals: The Political Genius of Abraham Lincoln* (New York: Simon and Schuster, 2005); Dick Morris, *The New Prince* (Los Angeles: Renaissance, 1999).

2. Willa Marie Bruce, *Problem Employee Management: Proactive Strategies for Human Resource Managers* (New York: Quorum, 1990); David D. Burns, *Feeling Good: The New Mood Therapy* (New York: Signet, 1980); Michael S. Dobson and Deborah S. Dobson, *Enlightened Office Politics: Understanding, Coping with, and Winning the Game—without Losing Your Soul* (New York: AMACOM, American Management Association, 2001); Andrew DuBrin, *Winning Office Politics: DuBrin's Guide for the '90s* (Englewood Cliffs, New Jersey: Prentice Hall, 1990); Roger Fisher, William Ury, and Bruce Patton, *Getting to Yes: Negotiating Agreement without Giving In*, 2nd ed. (New York Penguin, 1991); Roy H. Lubit, *Coping with Toxic Managers, Subordinates . . . And Other Difficult People* (Upper Saddle River, New Jersey: Pearson, 2004); Blaine Pardoe, *Cubicle Warfare: Self-Defense Strategies for Today's Hypercompetitive Workplace* (Rocklin, California: Prima Press, 1997); Marilyn Pincus, *Managing Difficult People: A Survival Guide for Handling Any Employee* (Avon, Massachusetts: Adams Media, 2004).

3. Linda Lear, *Rachel Carson: Witness for Nature* (New York: Henry Holt, 1997).

4. United States Senate, *June 9, 1954. Have You No Sense of Decency?* (2005 [cited April 20, 2005]); available from http://www.senate.gov/artandhistory/history/minute/Have_you_no_sense_of_decency.htm.

5. John Douglas and Mark Olshaker, *Mindhunter: Inside the FBI's Elite Serial Crime Unit* (New York: Pocket Books, 1995).

6. Lear, *Rachel Carson*.

7. "Feud in the Palmettos," *Time* 55, no. 14 (1950); "Fireworks in Florida," *Newsweek*, no. 15 (1950).

8. "Anything Goes," *Time* 55 no. 16 (1950); Robert Yoon, *Size Matters: A Little Knowledge Can Be a Dangerous Thing* (2005 [cited April 5, 2005]); available from http://www.ksg.harvard.edu/citizen/Oct04/YOON1004.HTM.

9. Jack Hurst, *Nathan Bedford Forrest: A Biography* (New York: Knopf, 1993).

10. Marc Reisner, *Cadillac Desert: The American West and Its Disappearing Water* (New York: Penguin, 1993).

11. *Simpson's Contemporary Quotations*, compiled by James B. Simpson. (Houghton Mifflin, 1988).

12. Predictions reported in *Audience Dialog. Hopelessly Wrong Predictions* (2005 [cited October 29, 2005]); available from http://www.audiencedialogue.org/predict.html; Chris Morgan and David Langford, *Facts and Fallacies: A Book of Definitive Mistakes and Misguided Predictions* (New York: St. Martin's Press, 1981); David Wallechinsky, Amy Wallace, and Irving Wallace, *The Book of Predictions* (New York: William Morrow, 1980).

13. Paul Ekman, Maureen O'Sullivan, and Mark Frank, "A Few Can Catch a Liar," *Psychological Science* May (1999).

14. Fisher, Ury, and Patton, *Getting to Yes*.

15. Jack Trimarco, *Fake Out: Jack's Tips* (2005 [cited January 15, 2006]); available from http://www.courttv.com/onair/shows/fake_out/bio_trimarco.html.

16. Fisher, Ury, and Patton, *Getting to Yes*.

17. Carville and Begala, *Buck up, Suck Up*; Morris, *New Prince*.

18. Niccolo Machiavelli, *The Prince*, ed. Daniel Donno, trans. Daniel Donno, Bantam Classic Edition, March 1981 ed. (New York: Bantam, 1513).

19. Brady, *Bad Boy*.

Chapter 11

1. "The Cities: The Price of Optimism," *Time*, August 1, 1969.

2. Gordon Young, "Pollution, Threat to Man's Only Home," *National Geographic*, December 1970.

INDEX

ABOUT THE AUTHOR

SCOTT BONAR is an Associate Professor of Natural Resources at the University of Arizona and Leader of the USGS Arizona Cooperative Fish and Wildlife Research Unit. He has conducted award-winning natural resources work in natural resources programs of state and federal government, universities, and private industry for over twenty-three years, authoring numerous publications, and supervising over eighty employees. In 2006, his unit won the highest award in the nation for relations with cooperators. He holds a B.S. degree in Science Education from the University of Evansville and a Ph.D. in Fisheries Science from the University of Washington. He lives in Tucson, Arizona, and spends much of his time explaining why the desert Southwest is a wonderful place for a fish biologist.